By the first days of 1944 the Axis forces operating on the Eastern Front had been driven back from much of the Soviet territory that they had gained during the invasion of June 1941 and the advances of 1942. In the north the Red Army was threatening the Baltic States while in the south Russian soldiers were pressing into Romania, an Axis ally.

The disaster at Stalingrad had hurt the Germans, although not fatally, but the repulse of the massive offensive effort made during the summer of 1943, which culminated in the battles around Kursk, had exhausted many of the Wehrmacht's best divisions. For the foreseeable future German units would be limited to defensive operations with local counterattacks employed to retake lost positions. The role of the remaining armoured formations in plugging any gaps in the line would be crucial. It should also be remembered that the Wehrmacht was at this time conducting a defensive campaign in Italy and had one eye on France where the coming year was sure to see an invasion by the Western Allies.

Anticipating that the Russians' main effort in 1944 would come in the southern sector of the front, securing a route to the Vistula and from there to Berlin, the Germans concentrated most of their armoured reserves with Heeresgruppe Süd. But when the Soviets finally attacked on 22 June 1944 (1) the main weight of the assault fell on Heeresgruppe Mitte and by late August many of the divisions on this sector had simply ceased to exist.

Codenamed Operation Bagration, this huge undertaking in fact comprised ten separate offensives, each involving more troops that took part in the Normandy landings. From this time the Germans could conduct fighting withdrawals at best, taking the offensive just once, in northern Hungary, before the end of the war in May 1945. It is against the backdrop of these momentous events that this second instalment of the Sturmgeschütz III story is set.

As explained in *Tankcraft 19: StuG III & IV, German Army, Waffen-SS and Luftwaffe, Western Front 1944-1945* the intended role of the Sturmgeschütz was to clear obstacles such as bunkers and pillboxes in support of infantry attacks. But necessity dictated that they were often employed as tank killers and the appearance of the long-barrelled 7.5cm L/48 gun in the spring of 1942 went some way towards cementing this position. As the war dragged on they were increasingly issued to replace tank destroyers and even tanks. In 1943 the concept of Gemischte, or mixed, Panzer battalions was introduced and it was also at this time that the first steps were taken to supply the Panzerjäger battalions of each infantry division with a company of assault guns. These developments are explained further in the charts on pages 61 to 63.

The principal model in service during 1944 was the Sturmgeschütz III ausf G, assembled by Alkett and MIAG, and the Sturmhaubitze 42, all of which were built at the Alkett plant in Berlin, and it is these variants which features most heavily in the following pages. Older models did, however, soldier on until the end of the war and a number of these are depicted in the Camouflage & Markings section.

The figures given here regarding vehicle strengths and allocations are largely based from contemporary reports. Documents such as the allocation lists, compiled by the Heereszeugamt (2) are remarkably accurate right up to the last days of the war. But the reader should note that in many unit reports specific models of vehicles are not always given and they can in fact be misleading. For example the word Sturmgeschütz is sometimes used without the qualifying suffixes of III or IV, often entered merely as 'StuG' (3). Where I have encountered any obvious anomalies or discrepancies I have endeavoured to highlight them.

On the following pages I have listed those units which served on the Eastern Front and were equipped with the Sturmgeschütz III and Sturmhaubitze 42 and while I have endeavoured to provide the reader with a clear picture of where those units operated and the equipment that they used, it should be noted that many were large and complex organisations that served from the very first days of the conflict and I have, of necessity, largely restricted the narrative to 1944.

I have deliberately omitted the Sturmgeschütz IV from this book as the story of the development of that vehicle and its employment in the East is worthy of its own volume.

It is an indication of the importance of the Sturmgeschütz III to the German war effort that the manufacture of an assault gun based on the Pzkpfw IV chassis would probably never have been contemplated if the Alkett works in Berlin had not been so comprehensively bombed in November 1943, affecting production for up to six months.

Notes

1. Commemorating the German invasion of 1941.
2. In many instances I have used the term 'allocation list' here and throughout this book in place of the German description Zuhführungslist purely as a matter of convenience and to spare the reader.
3. Similarly, it is apparent that in some instances the term has been used purely in a generic sense, employed as a form of shorthand to describe vehicles with a similar mission such as tank destroyers or self-propelled anti-tank guns.

From the earliest days of the Russian Campaign the Axis armies in the East were controlled by large Army Groups as shown here: Heersgruppe Nord (A); Heeresgruppe Mitte (B),and; Heeresgruppe Süd (C). Heeresgruppe Mitte was largely destroyed in the summer battles of 1944 and although it was rebuilt to some extent it could not hope to hold the extensive front line that had been created by the Soviet advance. The divisions of Heeresgruppe Mitte now operated between northern Poland and southern Lithuania (D). Heeresgruppe Nord was now restricted to the Baltic States and was eventually cut off there (E). The southern half of the front was now controlled by Heeresgruppe Nordukraine (F) and Heeresgruppe Südukraine (G). The first broken line (1) indicates the front line as it was on the evening of 21 June 1944 just prior to the commencement of Operation Bagration, the massive Soviet summer offensive. By mid-August 1944 the Red Army had advanced to the Vistula (2) and the second broken line (3) shows the front as it was on the last day of the year. The rapid expansion in the south was due in large part to the surrender of Romania and Bulgaria.

From their inception the Sturmgeschütz units were controlled by the artillery and, as opposed to those units attached to Panzer and Panzergrenadier divisions, remained under the direction of the Generalinspekteur der Artillerie. Their companies were therefore referred to as batteries, as they are throughout this book, and in the early units comprised six vehicles which was roughly the manpower equivalent of a normal artillery battery. By the beginning of the Russian campaign most of the batteries had been grouped into battalions referred to as Sturmgeschütz-Abteilungen and by the spring of 1942 each battalion was authorised twenty-eight assault guns. In November 1942, the battalions underwent another organisational change with each battery commander receiving their own assault gun, bringing the total number to thirty-one, and many units retained this establishment until the end of the war.

In mid-1943 it was proposed that the Panzergrenadier divisions that were being formed should contain an assault gun battalion and at the same time the concept of Gemischte, or mixed, tank battalions was introduced. In addition, the Panzerjäger battalions of infantry divisions were increasingly allocated a company of assault guns and, rather confusingly, these were referred to as Sturmgeschütz-Abteilungen. Vehicles assigned to Panzer, Panzergrenadier and Panzerjäger units were controlled by the Generalinspekteur der Panzertruppen and as early as October 1943 approximately 48 percent of Sturmgeschütz production was allocated to these formations.

In February 1944 it was decided that all independent battalions equipped with the Sturmgeschütz were to be renamed Sturmgeschütz-Brigaden. The Panzerjäger companies and assault gun units attached to divisions would henceforth be referred to as Sturmgeschütz-Abteilungen. In the same month, two new organisational changes were authorised for the independent assault gun battalions which were intended to provide each with a Sturmgeschütz-begleit battery, made up of motorised infantry and later referred to as a Begleitgrenadier-Batterie, and a Begleitpanzer battery equipped with Panzer II tanks. A final name change in June 1944 saw the Sturmgeschütz-Brigaden renamed as Heeres-Sturmgeschütz-Brigaden and the battalions that had received the Begleit companies were now referred to as Heeres-Sturmartillerie-Brigaden. These units are examined in detail on pages 61 and 62. As the war dragged on, assault guns were increasingly issued as replacements for tank destroyers and tanks. Their effectiveness, the relative ease with which they could be produced and their low cost to the German economy ensured their popularity with government and army planners.

Listed below are the units that were equipped with the Sturmgeschütz III assault gun that served on the Eastern Front during 1944. To avoid unnecessary repetition I have used the abbrevations StuG-Abt for Sturmgeschütz-Abteilung, StuG-Brig for Sturmgeschütz-Brigade, HStuG-Brig for Heeres-Sturmgeschütz-Brigade, HStuArt-Brig for Heeres-Sturmartillerie-Brigade, PzJg-Abt for Panzerjäger-Abteilung, StuH 42 for Sturmhaubitze 42 and of course StuG III for Sturmgeschütz III. Other abbreviations are explained throughout the text.

StuG-Abt 177. Formed in August 1941 at Jüterbog, by early 1943 the unit had been totally destroyed but was rebuilt in early 1944. In February of the same year, made up of just two companies, the battalion was attached to 3.Kavallerie-Brigade and in August was reorganised as PzJg-Abt 69. Photographic eveidence shows that at least one StuG IV assault gun was on hand during the summer of 1944.

StuG-Abt 184. Formed in August 1940, this unit was almost completely wiped out in the fighting for the Demyansk Pocket in early 1943 after which the survivors were sent to Estonia to refit. In February 1944 the battalion was renamed StuG-Brig 184 and took part in the fighting around Pskov, north of Ostrow in Russia, and the general withdrawal into southern Latvia.

In June 1944 the battalion was renamed H-StuG-Brig 184 and although it was withdrawn to rest and refit for much of July, reported in September that just six StuH 42 and four StuG III were on hand, one of the latter being an early model equipped with the 7.5cm L/24 gun. In late 1944 the battalion lost all its equipment fighting in the Kurland Pocket (1).

StuG-Abt 185. Formed in August 1940, the battalion fought on the Eastern Front from the invasion of the Soviet Union in July 1941 until the war's end in May 1945. In June 1944 the battalion was renamed StuG-Brig 185 and less than a month later underwent another name change to become HStuG-Brig 185. During the first week of August the battalion reported that seventeen StuG III and five StuH 42 were combat-ready. By October the battalion was under the control of XX.Armeekorps and withdrew into East Prussia.

StuG-Abt 189. Formed in July 1941, this battalion served with various units of Heeresgruppe (Hgr) Mitte until late 1942, when it was attached to 78.Sturm-Division (2).

Like many assault gun battalions this unit underwent a name change in June 1944 to become StuG-Brig 189 and just over a month later to HStuG-Brig 189. At the end of July 1944, the battalion was detached from 78.Sturm-Division and ordered to Magdeburg in Germany to refit. But before the refitting process was complete the battalion, less its third battery, was sent to Mielau in Poland and reorganised as PzJg-Abt 70 while some elements were attached to 4.Kavallerie-Brigade.

The third battery remained in Magdeburg and provided the cadre for 3.Batterie, HStuG-Brig 244.

StuG-Abt 190. Formed in October 1940, this unit served in the Balkans and Greece before taking part in the invasion of Russia in July 1941. Re-named leichte-StuG-Brig 190 in January 1944, the battalion took part in the fighting around the city of Kovel in Poland, supporting SS-Panzer-Regiment 5 (3). By May 1944, the battalion was in the Mogilev area attached to

Notes

1. An order dated 15 December 1944 directed that this battalion should be reorganised with a Begleitgrenadier battery but is not clear if this was ever carried out.

2. Interestingly, the battalion was originally attached to the artillery regiment but later came under the direct control of the division's commander.

3. The leichte, or light, prefix was added in April 1943 and would seem to indicate that this unit's organisation was somehow different to the other independent battalions but I have been unable to ascertain if this is correct. Only StuG-Brig 600 (see below) was similarly named.

4.Armee and in September reported that twenty-two StuG III and four StuH 42 were operational while two StuG III and one StuH 42 were in repair.

In November 1944 the battalion was ordered to Germany for refitting and returned to the front near Danzig as part of 2.Armee in January 1945.

StuG-Abt 191. Formed in October 1940, this unit served in the Balkans and Greece before taking part in Operation Barbarossa in July 1941. In June 1944 the battalion, by now renamed StuG-Brig 191, was refitting at Warthengau in Germany and in the following month was reinforced by a fourth battery, formed from 4.Batterie, HStuG-Brig 201, and was transferred to eastern Yugoslavia as part of 2.Armee.

This battalion was one of the few StuG units that were authorised to contain forty-five assault guns. In September 1944 the battalion's second battery was removed and transferred to Lehr-Brigade II at Altengrabow.

StuG-Abt 201. Formed in March 1941, this unit was almost completely destroyed in the Stalingrad battles and reformed in May 1943, spending most of that year in Greece. In March 1944, by now renamed StuG-Brig 201, the battalion was part of Kampfgruppe Hildebrandt, fighting with XXII.Gerbirgs-Armeekorps in Yugoslavia.

In October 1944, the battalion was attached to 4.Panzerarmee and by early 1945, the few survivors were merged with StuG-Brig 210.

StuG-Abt 202. Formed in September 1941, this battalion took part in some of the heaviest fighting on the Eastern Front until it was almost completely destroyed in the battles for the Cherkassy-Korsun Pocket in early 1944.

Renamed StuG-Brig 202, the battalion was ordered to Ziegenhals, near Danzig in Poland, to refit. By August, the battalion was again at the front fighting in the Kurland Pocket. Here the battalion took part in the battles for Ösel Island, modern day Saaremaa in the Gulf of Riga, and around Tukums in Latvia.

An order dated 15 December 1944 directed that the battalion should be reorganised with a Begleitgrenadier battery but it is unclear if the additional unit was ever received (1).

StuG-Abt 203. Formed in February 1941, this battalion fought until May 1945 on the Eastern Front. By February 1944, it had been re-named StuG-Brig 203 and was assigned to LII.Armeekorps, part of Hgr Süd. By October the battalion was under the command of 4.Armee and ended the war with 1.Panzerarmee fighting in Silesia and eastern Germany.

StuG-Abt 209. Formed in December 1941, this battalion spent the early months of 1944 refitting in Poland and was renamed StuG-Brig 209 in February. By the first week of July, the battalion had twenty-two StuG III and nine StuH 42 on hand and was transferred back to the front as part of 2.Armee.

Notes

1. Strength reports from the first days of 1945 suggest that the battalion may have been organised on the forty-five assault gun establishment.

Photographed during the spring of 1944, this MIAG-built Sturmgeschütz III ausf G is fitted with the Schwingschürzen, or swinging side skirts, pioneered by Sturmgeschütz-Abteilung 190. Note the method by which the plates are attached to the trackguard. Further examples are shown and discussed in the Camouflage & Markings section. The pattern of Zimmerit which identifies the manufacturer is clearly visible on the front fender.

An Alkett-built Sturmgeschütz III ausf G of Sturmgeschütz-Brigade 202 photographed during the summer of 1944. Note the armoured engine covers fitted to the glacis and the concrete reinforcement on the superstructure front, both of which were common modifications within this unit. Note also that lower edge of the first plate of the Schürzen has been cut away to form a curve. All the battalion's assault guns were named, with the name displayed on the gun mantlet as seen here.

StuG-Abt 210. Formed in May 1941, this unit served exclusively on the Eastern Front. In February 1944, the battalion was renamed StuG-Brig 210 and spent most of that year subordinated to 1st Hungarian Army fighting in Ukraine.

In September 1944 the battalion reported that thirteen StuG III and six StuH 42 were operational with a number of additional vehicles in repair.

In November the battalion was attached to LVI.Panzerkorps of 4.Panzerarmee.

StuG-Abt 226. Formed in early 1941, this battalion was renamed StuG-Brig 226 in February 1944 and remained with Hgr Nord throughout that year. In November 1944, the battalion was completely re-equipped with Panzer IV/70 tank destroyers and a fourth, Begleit-Batterie, was authorised although it is unclear if this was ever received (1).

StuG-Abt 228. Formed in December 1942, this battalion was continuously active on the Eastern Front and by summer 1944 had been renamed StuG-Brig 228 and attached to the 1st Romanian Infantry Division under the command of Hgr Südukraine. In September 1944 the battalion reported that it had no serviceable vehicles on hand but a handwritten note on the December allocation list suggests that four StuG III were sent in that month.

StuG-Abt 232. Formed in October 1942, this battalion spent the early months of 1944 refitting and as part of an operational reserve force for 6.Armee. In February the battalion was renamed Stug-Brig 232 and in July was attached to 3.Panzerarmee. In September, the battalion came under the orders of IX.Armeekorps and reported that eighteen StuG III and seven StuH 42 were on hand at that time.

StuG-Abt 236. This unit was formed in March 1943, from 3.Batterie, StuG-Abt 189 and StuG-Ersatz-Abt 300. In early February 1944 the battalion was renamed StuG-Brig 236 and in the same month a Begleitgrenadier-Batterie and a Begleitpanzer-Batterie, numbered as the fourth and fifth batteries, were authorised but it was some time before these units completed their training and reached their parent formation.

In June the battalion was renamed HStuArt-Brig 236. In September, as part of Hgr Südukraine, the battalion was almost completely destroyed in the fighting around the town of Jassy in Romania near the present-day border with Moldava but within a month had been reformed around a cadre of eighty-seven survivors.

In November 1944 the Begleitpanzer-Batterie was officially withdrawn but a number of the Pzkpfw II tanks were still on hand as late as March 1945.

StuG-Abt 237. Formed in mid-1943, this battalion served in Russia under the command of Hgr Mitte until July 1944, when it was withdrawn from the front and ordered to Möckern, near Magdeburg in Germany (2).

Notes

1. There is also some evidence that at this time the battalion was renamed HStuArt-Brig 226.
2. Here, with 4.Batterie, Sturmpanzer-Abteilung 216, the battalion was used to form Sturmpanzer-Abteilung 219 equipped with the Sturmpanzer IV.

Notes

1. They were numbered as the fourth and fifth batteries and the latter was equipped with Pzkpfw II tanks.

StuG-Abt 239. Formed in July 1943, this unit lost all its heavy equipment in the Cherkassy-Korsun Pocket in February 1944. Rebuilt almost immediately as HStuArt-Brig 239, the battalion received the Begleitgrenadier-Batterie and the Begleitpanzer-Batterie that were originally authorised for StuG-Brig 905 (1).

In August 1944, the battalion suffered heavily in the fighting for Romania, losing most of the second and third batteries. In September, the battalion was reformed from the survivors of 3.Batterie and parts of StuG-Brig 184, StuG-Brig 189 and StuG-Brig 236. The Panzer battery was officially withdrawn in November but there is some evidence that a number of the vehicles were retained.

StuG-Abt 243. Formed in May 1941, this unit fought almost exclusively on the Eastern Front and for a time was attached to 24.Panzer-Division. In early 1943 the battalion was severely mauled in the Stalingrad battles and was withdrawn from the front. It was reformed in February 1944 and renamed StuG-Brig 243. In June 1944, the battalion was in Ukraine as part of 6.Armee and remained there until December, when it was transferred to the West.

StuG-Abt 244. Formed in June 1941, and renamed StuG-Brig 244 in February 1944, this battalion was part of Hgr Mitte when the massive Soviet summer offensive began. The battalion was destroyed in the fighting around Bobruisk, south-east of Minsk. Rebuilt in Holland in October the battalion remained in the West until the war's end.

StuG-Abt 245. Formed in June 1941, this unit was renamed StuG-Brig 245 in February 1944. The battalion was subordinated to 3.Panzerarmee in June and had on hand twenty-two StuG III and six StuH 42, a small number of each in need of repair. But by August the battalion's vehicle strength had been reduced so dramatically that it was disbanded, with the survivors being absorbed by StuG-Brig 667.

StuG-Abt 249. Formed in January 1942, this battalion was part of an ad hoc force co-ordinated by 1.Panzerarmee which took part in the relief of the Cherkassy-Korsun Pocket in early 1944. Renamed StuG-Brig 249 in February, the battalion was completely destroyed in the fighting for Brody, between Lviv and Rivne in present day Ukraine, in July.

By October, the battalion had been rebuilt and was back on the Eastern Front with XXIII.Armeekorps. In late 1944 or early 1945, the battalion was again withdrawn from the front and renamed HStuArt-Brig 249.

StuG-Abt 259. Formed in June 1943, the battalion was renamed StuG-Brig 259 in February 1944. It was attached to 6.Armee when the Soviet summer offensive began in June 1944 and in the following month the battalion was authorised to have a full complement of forty-five assault guns. At the same time the battalion was fighting around Bialystok in Poland, north-east of Warsaw. In September 1944 the battalion reported that twenty-nine StuG III and eight StuH 42 were operational while a single Sturmhaubitze was in repair.

Elements of Heeres-Sturmartillerie-Brigade 239 on their way to the front. Almost a full battery of Sturmgeschütz III ausf G assault guns are visible in the background, all fitted with Schürzen, while at least two Pzkpfw II tanks of the battalion's 5.(Panzerbegleit)-Batterie are closest to the camera. The presence of the latter date this image to some time after May 1944. Note the three-colour camouflage pattern and the small Hakenkreuz flag in front of the commander's cupola of the nearest vehicle.

Photographed during the last weeks of 1943 or very early in 1944, this Sturmgeschütz III ausf G is from 1.Batterie, Sturmgeschütz-Brigade 286. The officer in the background is Oberleutnant, later Hauptmann, Paul Dahms who commanded the batterie and subsequently the brigade. Note the loader's MG 42 and the enlarged aperture of the gun's protective shield. The barrel of the 7.5cm main gun appears to have been left in its dark lacquer coating which was quite common.

StuG-Abt 261. Formed in July 1943, this unit was fighting near Kirovograd with XI.Armeekorps as part of Hgr Süd in early 1944. By August, the battalion had been transferred to Hgr Nord where it remained until 1945.

StuG-Abt 270. Formed in October 1942, this battalion was temporarily attached to 1.Skijäger-Brigade in January 1944. In June, the brigade was expanded to a full division and the assault gun battalion was renamed StuG-Brig 270 at the same time.

By September it had been decided that the battalion should be permanently assigned and it was renamed once again as PzJg-Abt 270 which was almost immediately changed to PzJg-Abt 152. In January 1945, the battalion reported that twelve StuG III and two StuH 42 were on hand.

StuG-Abt 276. Formed in June 1943, this unit lost almost all its equipment in the battles for the Kamyenets-Podolskyi Pocket in Ukraine in April 1944, although most of the personnel survived. In May, the battalion was rebuilt and by August it was back at the front, fighting in Lithuania and East Prussia. The battalion had been renamed StuG-Brig 276 in February 1944.

StuG-Abt 277. Formed in early 1943, this battalion served on the Eastern Front until December of the same year when it was withdrawn to Germany to be completely re-equipped. The battalion was renamed StuG-Brig 277 in February 1944 and returned to the front in the following month but apparently took no part in the fighting until June when it was transferred to 4.Panzerarmee which was operating around Cholm in Poland. In September 1944 the battalion reported that sixteen fully operational StuG III and seven StuH 42 were available while two StuG III and one StuH 42 were in repair.

StuG-Abt 278. Formed in August 1943, little is known about this unit. During the summer battles, the battalion was attached to the 3rd Romanian Army of Hgr Süd and was almost completely destroyed during the fighting in August. From September until the end of the year, the battalion was refitting in Burg in Germany and in January 1945, was absorbed by StuG-Brig 232.

StuG-Abt 279. Formed in July 1943, this battalion was severely mauled in the battles in the Crimea in May 1944 and withdrawn to Germany to be rebuilt.

This battalion was one of just four StuG units to be officially authorised to contain forty-five assault guns but in reality this reorganisation may never have been carried out as in late August 1944 the battalion reported that just twenty StuG III and ten StuH 42 on hand (1).

StuG-Abt 280. Formed in August 1943, the battalion was renamed StuG-Brig 280 in February 1944. The battalion was engaged in the fierce defensive battles around Kiev in Ukraine and during April 1944 took part in the attempt to relieve Tarnopol, suffering heavy casualties in the process. Ordered to Denmark to refit, the battalion spent the rest of the war on the Western Front.

Notes

1. In addition, I have been able to find any further allocation.

Notes

1. During late 1944 or early 1945, the battalion received three Panzer IV/70(A) tank destroyers.

2. At least one authoritative source states that the battalion arrived in northern Russia in January without any vehicles.

StuG-Abt 281. Formed in October 1943, the battalion was re-named StuG-Brig 281 in February 1944. In June, the battalion was attached to 3.Panzerarmee of Hgr Mitte and was located near Vitebsk with twenty-nine assault guns on hand. In September, the battalion was disbanded and used to form Artillerie-Pak-Abteilung 1052, a towed anti-tank gun unit which was completely destroyed in France in September 1944.

StuG-Abt 286. Formed in August 1943, the battalion spent the last months of the year in France until it was transferred to the East in late December. Renamed StuG-Brig 286 in February 1944, the battalion was under the direct orders of Hgr Mitte throughout 1944 and took part in the fighting in Ukraine. In September the battalion reported that twenty-six fully operational StuG III were on hand.

StuG-Abt 300. Formed in October 1943, the battalion trained in France until December when it was transferred to the Eastern Front. In February 1944 the battalion was renamed StuG-Brig 300 and by April was under the command of Hgr Nordukraine. In September 1944 the battalion reported that fourteen operational StuG III were on hand with nine StuG III and a single StuH 42 in repair (1).

StuG-Abt 301. Formed in October 1943, partly from the remnants of StuG-Abt 243 which had been destroyed at Stalingrad, the battalion was transferred to the East in February 1944 and re-named StuG-Brig 301. The first battery was detached from the battalion and destroyed at Tarnopol and the remainder of the brigade was assigned to 1.Panzerarmee. In September 1944, the battalion was transferred to Slovakia and in the following month was moved to Krakow in Poland for refitting. A planned fourth battery was not achieved and the battalion returned to the front in January 1945.

StuG-Abt 303. Formed in October 1943, the battalion spent the remainder of the year training in France and before leaving for the East was issued with thirty StuG III and twelve StuH 42 (2). In June 1944, the battalion was attached to Armeeabteilung Narwa and during that month was transferred to Finland in an effort to support Germany's ally.

Before the battalion left Russia authorisation had been received for a fourth battery but this was not realised until the autumn when Stug-Abt 303 was already operating in Finland. In September 1944, after being withdrawn from Finland, the battalion was attached to Panzerverband von Lauchert to reinforce 3.Panzerarmee. In October 1944, the battalion was renamed HStuArt-Brig 303 but it is unlikely that the proposed Begleitgrenadier-Batterie was ever formed. In November 1944, the battalion was assigned to 6. Armee in Hungary.

StuG-Abt 311. Formed in November 1943, the battalion was renamed StuG-Brig 311 in February 1944 and sent to the Eastern Front in the following month. It was immediately in action in the Tarnopol

Photographed in Finland during the summer of 1944 this Sturmhaubitze 42 is from 1.Batterie, Sturmgeschütz-Brigade 303, as the battalion was referred to at the time. Note the large armoured cover over the driver's visor and heavy logs fitted between the Schürzen and the superstructure sides. Both of these modifications may have been due to the influence of the Finnish StuG units

area where it was almost completely destroyed. The survivors were withdrawn some 60 kilometres to the west to Berezhany were the battalion was reformed and was once again at the front in July (1).

The battalion ended the war at in the besieged city of Wroclaw in Poland with a single surviving assault gun.

StuG-Abt 322. Formed in November 1943, possibly around a cadre from 3.Batterie, StuG-Abt 184, by February 1944 this unit had been renamed StuG-Brig 322. It was originally planned to equip this battalion with a Begleitgrenadier-Batterie and a Begleitpanzer-Batterie but the idea was very quickly dropped.

Rushed to the front in March, the battalion took part the fighting for Tarnopol and the battles around Brody, near Lviv in present-day Ukraine. In October 1944, the battalion was attached to LVI.Panzerkorps of 4.Panzerarme where it remained until January 1945 when it was disbanded.

StuG-Abt 325. Formed in April 1943 from a cadre of 1.Batterie, StuG-Abt 912, the battalion was renamed StuG-Brig 325 and sent to the front in April where it took part in the fighting around Jassy in Romania in support of the Romanian Guards Infantry Division.

In August, the battalion was attached to Armeegruppe Wöhler which at that time contained units of the German 8.Armee and the Romanian 4th Army and was fighting a major withdrawal action in southern Ukraine, eventually retreating into Romania.

The battalion was almost completely destroyed in the fighting here and by December the survivors, who had escaped on foot, were regrouped in Hungary where they were able to form a single battery.

StuG-Abt 393. Formed in March 1944, the battalion was assigned to Hgr Nord in June or July 1944, and stationed around Daugvapils in Latvia. In August the battalion was subordinated to 18.Armee and the following month the first battery was detached and placed under the command of Armeeabteilung Narwa, a large ad hoc formation made up of Army and Waffen-SS units.

At the end of 1944, the battalion was renamed HStuArt-Brig 393 and an order dated 15 December 1944 directed that the battalion was to receive a Begleitgrenadier battery. The battalion ended the war fighting in the Kurland Pocket.

StuG-Abt 395. Formed in May 1944, this battalion saw limited service on the Eastern Front until July when it was disbanded and used to form StuG-Abt 1550, 1551, 1552, 1553, 1558 and 1559.

StuG-Abt 600. Formed in July 1940, this unit was almost completely destroyed in the fighting of late 1943 and early 1944 (2). The battalion was rebuilt and sent back to the front in November 1944, where its third battery, with nine assault guns, was attached to 4.Panzer-Division.

In the autumn of 1944 the battalion was renamed HStuArt-Brig 600 but it is unclear if the two proposed Begleit batteries were ever received although strength reports submitted in early 1945 suggest that this unit may have been organised on the forty-five assault gun establishment.

StuG-Abt 667. Formed in June 1942, the battalion was renamed StuG-Brig 667 in February 1944. In the same month a Begleitgrenadier-Batterie and two Begleitpanzer-Batterien, numbered as the fourth, fifth and sixth batteries, were authorised. This was the only assault gun unit to receive two Panzer batteries.

But it was not until May or June that these units completed their training and reached their parent formation and by that time the battalion had been renamed HStuArt-Brig 667. The battalion was almost completely wiped out during the massive Russian summer offensive in July 1944, losing both tank batteries which were never replaced.

During the autumn of 1944 the battalion was rebuilt, partly from the remnants of StuG-Brig 245, and spent the remainder of the war on the Western Front.

StuG-Batterie 741. Formed in Finland in January 1943, this unit was raised from volunteers of the various independent tank formations that had been stationed there. The battalion was caught in the encirclement of 1.Panzerarmee in April 1944 and was so badly mauled that it was withdrawn to Saarbrücken in Germany and disbanded in July.

Most of the surviving crews and vehicles were absorbed by StuG-Brig 394 with the surplus personnel being allocated to the replacement pool.

StuG-Abt 904. Formed in December 1942, this battalion was renamed StuG-Brig 904 in February 1944. The battalion was assigned to 2.Armee in June 1944 reporting that only fourteen StuG III assault guns were on hand.

In July the battalion was attached to 4.Panzer-Division that was itself part of Gruppe Harteneck, a large ad-hoc formation commanded by General Gustav von Harteneck, which also contained 4.Kavallerie-Brigade and the remnants of 29.Infanterie-Division. The battalion remained with 2.Armee until 1945 when it disappears from the records.

Notes

1. Plans to equip this battalion with a Begleitgrenadier-Batterie and a Begleitpanzer-Batterie, numbered as the fourth and fifth batteries, were eventually abandoned. A number of authoritative sources state that this was the first unit to receive an allocation of StuG IV assault guns but this is a matter of some debate.
2. Also referred to as leichte StuG-Abt 600. See also Stug-Abt 190.

StuG-Abt 905. Formed in December 1942, the battalion was renamed HStuG-Brig 905 in February 1944. In the same month a Begleitgrenadier-Batterie and a Begleitpanzer-Batterie, numbered as the fourth and fifth batteries, were authorised but for some reason the order was countermanded and the additional units went to StuG-Brig 239. In June the battalion was supporting the 4th Romanian Army and was destroyed in the fighting between Jassy and Kishinev in August 1944. Reformed later that year as HStuArt-Brig 905, the battalion spent the remainder of the war in the West.

StuG-Abt 909. Formed in January 1943, the battalion was renamed StuG-Brig 909 in February 1944. In June the battalion was directly subordinated to Hgr Nord and was able to report that it had twenty-one StuG III and eight StuH 42 on hand. The September 1944 strength return lists only the second battery as operational with just six StuG III and one StuH 42 in repair.

StuG-Abt 911. Formed in February 1943, the battalion was renamed StuG-Brig 911 in February 1944. In August 1944, the battalion was completely destroyed in the fighting in Romania and later reformed in Germany with three StuG batteries and one Begleitgrenadier battery and renamed HStuArt-Brig 911. In December 1944, the battalion was attached to the newly formed Führer-Grenadier-Brigade.

StuG-Abt 912. Formed in February 1943, the battalion was renamed StuG-Brig 912 in February 1944. This unit was subordinated to X.Armeekorps of 16.Armee during the early summer of 1944 but by September, was under the control of Gruppe General Kleffel, a corps-sized ad hoc unit which also contained 93.Infanterie-Division and a Waffen-SS division made up of Latvian volunteers. By December 1944, the battalion was attached to 11.Infanterie-Division in the Kurland Pocket and at about this time was renamed HSturArt-Brig 912. Strength reports from the first days of 1945 would suggest that this unit may have been organised on the forty-five assault gun establishment and an order dated 15 December 1944 did in fact authorise the addition of a Begleitgrenadier battery.

StuG-Abt Burg. Formed in July 1944 from 1.Lehr-Batterie of the StuG-Schule Burg, the battalion was almost immediately sent to the Eastern Front and spent most of the of the year with 4.Armee, initially with Kampfgruppe Weidling.

StuG-Brig I. Formed in July 1944, from elements of the StuG-Schule Burg, the battalion was originally established with four batteries, but the fourth was detached almost immediately and assigned to StuG-Brig 209. The remainder of the battalion was sent to Poland and assigned to VIII.Armeekorps and renamed StuG-Lehr-Brigade 920.

StuG-Lehr-Brig 920. Created by renaming StuG-Brig I in late 1944, the battalion was almost completely destroyed in the fighting between the Vistula and Oder rivers.

StuG-Brig Grossdeutschland (GD). Originally formed from 16.Kompanie (StuG), Infanterie-Regiment GD and parts of StuG-Abt 192, the battalion was made up of a headquarters staff and three assault gun companies. In late 1944 the battalion was detached from its parent division, Panzergrenadier-Division GD, and remained a separate entity until the end of the war, for the most part subordinated to Panzerkorps GD (1).

Notes

1. It may be that the original intention was to incorporate the assault guns into Panzergrenadier-Division Brandenburg as II.Abteilung of the division's Panzer regiment.

Two Sturmgeschütz III ausf G control vehicles of 3.Kompanie, Panzer-Abteilung (Funklenk) 302 photographed near Plac Teatralny in central Warsaw. The battalion lost two assault guns in an attack close to this location in Krakowskie Przedmiescie Steet on 23 August 1944. The elaborate camouflage scheme of random swirls was applied to all the vehicles of this company. This formation and its equipment are discussed in detail on page 63.

A Sturmgeschütz III of 9.Schwadron, Panzer-Regiment 24 photographed in northern Italy, shortly before the regiment moved to the Eastern Front. Forty-two vehicles were shipped to the regiment in May and June 1943 and this vehicle is almost certainly one of those. No further allocation was made and it is reasonably safe to assume that the three remaining assault guns reported as being available in March 1945 were also early model vehicles. This assault gun is fitted with the support rails for the Series I Schürzen, smoke grenade dischargers and rotatable cupola hatch. Note that the company number 943 on the superstructure side, rendered in red, is also just visible on the rear hull.

From early 1943 a number of Panzer battalions were reorganised as Gemischte, or mixed, units made up of tanks and assault guns. Only a small number of these Panzer-Sturmgeschütz-Abteilungen (Pz-StuG-Abt) were raised before the end of 1944 and they are listed here.

Panzer-Regiment 2. Attached to 16.Panzer-Division, the regiment was wiped out at Stalingrad and reformed in February and March 1943. In the following November III.Abteilung was converted to a Pz-StuG-Abt of three companies, each equipped with ten StuG III assault guns with a further vehicle controlled by the battalion headquarters.

On 31 May 1944 the division reported that nineteen StuG III were available, although seven of those were not considered combat ready, and in June a further seventeen new vehicles were received. By 1 July the battalion had nineteen serviceable assault guns on hand and during the same month received seventeen replacement vehicles, losing six in the same period.

Panzer-Regiment 23. A detailed organisational chart prepared for this regiment's parent formation, 23.Panzer-Division, dated 1 July 1944 shows I.Abteilung made up of three tank companies equipped with Pzkpfw IV tanks and a StuG company (1).

Panzer-Regiment 24. Attached to 24.Panzer-Division, the regiment was rebuilt in mid-1943 after being completely destroyed at Stalingrad. The regiment's III.Abteilung was converted to a Pz-StuG-Abt made up of two companies of StuG III assault guns, numbered as 9.Schwadron and 11.Schwadron (2), and two squadrons of Pzkpfw IV tanks. On 31 May 1944 the regiment reported that sixteen assault guns were on hand, although only nine of those were combat ready, while a further four were received on the last day of June (3).

Panzer-Regiment 36. Attached to 14.Panzer-Division, the regiment's III.Abteilung was reorganised as a Pz-StuG-Abt in mid-1943 with two companies of Pzkpfw IV tanks and two companies of StuG III assault guns, the latter numbered as 10.Kompanie and 12.Kompanie. On 31 May 1944 the regiment reported that just three assault guns were on hand but by the end of June fifteen fully-operational StuG III were available.

Panzer-Abteilung 5. The battalion was formed as a Pz-Stug-Abt with three companies of StuG III assault guns in August 1943 and at that time each company was equipped with fourteen assault guns. Originally assigned to 20.Panzergrenadier-Division, by October 1943 Panzer-Abteilung 5 was with 25.Panzergrenadier-Division. In early 1944 the division reported that forty-six assault guns were on hand, although five of these required some form of repair. By November the division had been reduced to a Kampfgruppe made up primarily from Panzergrenadier-Regiment 35 and PzJg-Abt 25.

Panzer-Abteilung 7. Attached to 10.Panzergrenadier-Division, the battalion had been formed as a Pz-StuG-Abt of three companies, each with fourteen

Notes

1. The only available strength return we have for this division is the 15 March 1945 report which shows seven operational assault guns on hand and three in repair. It is unlikely that these vehicles were allocated to the division's PzJg-Abt 128 which was known to have been equipped with Jagdpanzer IV tank destroyers.

2. As a former cavalry regiment this formation retained many of the traditions associated with that branch of the army including the reference to the companies as squadrons.

3. Interestingly, Panzer-Artillerie-Regiment 89, which was also attached to the division, was issued eight StuH 42 in March 1944. All eight vehicles were handed over to the regiment's 2.Batterie in the following month and remained with that unit until January 1945.

StuG III assault guns in August 1943.The division spent the winter of 1943-44 fighting in the Kremenchug area in present-day Moldova and was badly depleted during the Soviet summer offensive. The division was reformed as a Kampfgruppe in October and Pz-Abt 7 was once again re-equipped with StuG III assault guns and returned to the front.

Panzer-Abteilung 118. Attached to 18.Panzergrenadier-Division, this battalion was formed in September 1943 with three companies, each equipped with fourteen StuG III assault guns and reported on 31 May 1944 that forty-four assault guns were on hand. The division was largely destroyed in the summer battles with the surviving personnel used to form Panzer-Brigade 105. The division was rebuilt during November and December 1944 from various scratch units and Pz-Abt 118 was again equipped with a number of StuG III assault guns.

Panzer-Abteilung (Funklenk) 302. This unit is discussed in detail on page 63.

As one way of compensating for the greatly reduced manpower of the army's divisions it was planned that the PzJg-Abt of each should contain a company of assault guns made up of ten vehicles, either StuG III or IV assault guns or Jagdpanzer 38(t) tank destroyers (1). Usually, but not always, the Sturmgeschütz company was numbered separately from its parent battalion, the number arrived at by adding 1000 to the division's number, for example the assault gun company of PzJg-Abt 31 was referred to as StuG-Abt 1031.

PzJg-Abt 1. Attached to 1.Infanterie-Division, in late 1943 the battalion received an assault gun company which was referred to as StuG-Abt 1001. The battalion was originally allocated fourteen StuG III assault guns in October 1943 but in July 1944 a total of ten StuG IV vehicles were dispatched suggesting that all the StuG III had been lost.

PzJg-Abt 6. Attached to 6.Infanterie-Division, the battalion was allocated ten StuG III in January 1944 and these were assigned to 2.Kompanie which was referred to as StuG-Abt 1006. The division was completely destroyed in June 1944 but was later rebuilt, initially as 552.Grenadier-Division and later as 6.Grenadier-Division. The August 1944 allocation list shows that ten Stug III were assigned to the battalion (2) and the assault gun company was referred to as StuG-Abt 6. In October 1944 the division was again renamed 6.Volksgrenadier-Division.

PzJg-Abt 7. Attached to 7.Infanterie-Division, the battalion's 2.Kompanie was formed from parts of PzJg-Abt 215 and referred to as StuG-Abt 1007. Ten StuG III were shipped to the battalion in January 1944.

PzJg-Abt 11. Attached to 11.Infanterie-Division, the battalion was sent ten StuG IV assault guns in June 1944 although it is likely that this shipment was redirected to another, so far unknown, unit. Further evidence suggesting that the June shipment was not retained is the record of ten StuG III being sent in August (3).

PzJg-Abt 21. Attached to 21.Infanterie-Division, the battalion's assault gun company was referred to as StuG-Abt 1021. In September 1943 the battalion received fourteen StuG III but these were apparently lost in the withdrawal from the Leningrad Front and the fighting in the Baltic as the battalion was reported to be training on the Jagdpanzer 38 in October 1944.

PzJg-Abt 27. Attached to 17.Panzer-Division, the battalion had seven operational StuG III assault guns on hand in July 1944, losing all by the end of the month. The battalion was transferred to Germany to refit but probably never returned to its parent formation.

PzJg-Abt 28. Attached to 28.Jäger-Division, the battalion's assault gun company was referred to as StuG-Abt 1028.

PzJg-Abt 30. Attached to 30.Infanterie-Division, the battalion's assault gun company was referred to as StuG-Abt 1030. The battalion was allocated ten StuG IV assault guns in late April 1944 but these were almost certainly diverted to 57.Infanterie-Division (4). A further shipment of ten StuG IV was dispatched in May 1944 but these were subsequently lost or handed over to another unit as ten StuG III were shipped to the battalion on 1 October.

PzJg-Abt 31. In June 1944 the battalion's parent formation, 31.Infanterie-Division, was almost completely destroyed. In July the division was rebuilt and briefly known as 550.Grenadier-Division and a total of ten StuG III were assigned at that time (5). The PzJg-Abt contained only StuG-Abt 1031 and Flak-Kompanie 31 until February 1945 when the two companies were combined (6).

PzJg-Abt 35. Attached to 35.Infanterie-Division. The September 1944 allocation list shows that ten Stug III were allocated at that time. Confusingly, the division's strength reports show that the battalion was equipped with Marder III self-propelled guns in April 1944 and at least three were still on hand in early 1945.

PzJg-Abt 37. Attached to 1.Panzer-Division, the battalion's 2.Kompanie reported that twelve StuG III assault guns were on hand in July 1944 but only three remained by the end of the month. In December the company handed over all its surviving vehicles to 1.Kompanie and returned to Grafenwöhr in Germany.

Notes

1. An essential difference between these units and the assault gun brigades was that they were controlled by the Generalinspekteur der Panzertruppen.
2. The entry actually refers to 552.Grenadier-Division which was being rebuilt at that time but had in fact been renamed 6.Grenadier-Division in the previous month. A number of depleted divisions were rebuilt in late 1944 and given numbers in the 500 series as a temporary measure, later reverting to their previous titles.
3. In December 1944, a shipment of ten StuG IV assault guns was sent and this would suggest that all the StuG III vehicles had been lost.
4. Confusingly, the assault gun companies of the two divisions may have been exchanged with a subsequent change of name for both.
5. Before the end of July the division had been renamed 31.Grenadier-Division.
6. In some documents the battalion is referred to as StuG-Abt 31.

PzJg-Abt 45. In July 1944 a total of ten StuG III were allocated to 546.Grenadier-Division which was renamed 45.Infanterie-Division at about the same time (1). The PzJg-Abt was built from elements of Grenadier-Regiment 130 and Artillerie-Regiment 98 and the assault gun company was referred to as StuG-Abt 1045.

PzJg-Abt 69. See StuG-Abt 177 above.

PzJg-Abt 92. Attached to 20.Panzer-Division, the battalion operated a number of StuG III assault guns in late 1943 and early 1944. In May 1944 the battalion was transferred to the Panzerjäger-Schule at Mielau and completely re-equipped with Jagdpanzer IV tank destroyers.

PzJg-Abt 122. Attached to 122.Infanterie-Division, the battalion was sent ten StuG III assault guns in July 1944 for use by 2.Kompanie which was referred to as StuG-Abt 1122. The division was completely destroyed in the fighting for the Kurland Pocket

PzJg-Abt 126. Attached to 126.Infanterie-Division, the battalion was sent ten StuG IV assault guns in July 1944, for use by its 2.Kompanie which was referred to as StuG-Abt 1126. What became of these vehicles is unclear but on 19 December 1944 a total of ten StuG III were shipped to the division and these were reported as being on hand, together with one Bergepanzer III recovery vehicle, in the following month.

PzJg-Abt 132. Attached to 132.Infanterie-Division, the August 1944 allocation list shows that ten Stug III were assigned. Various reports from the division also indicate that a number of StuG IV assault guns were on hand.

PzJg-Abt 152. See StuG-Brig 270 above.

PzJg-Abt 158. Attached to 58.Infanterie-Division, the battalion was allocated ten StuG IV assault guns in June 1944 and six StuG III in the following August.

PzJg-Abt 169. Attached to 69.Infanterie-Division. The August 1944 allocation list shows that ten StuG III were allocated.

PzJg-Abt 173. Attached to 73.Infanterie-Division. The September 1944 allocation list shows that ten StuG III were assigned.

PzJg-Abt 187. Attached to 87.Infanterie-Division, the battalion received an assault gun company in April 1944 which was referred to as StuG-Abt 1187. At that time the battalion reported that six StuG III were on hand with the total rising to ten in July. In September 1944 the division reported, after weeks of fighting in Latvia, that just five StuG III were available.

PzJg-Abt 195. Attached to 95.Infanterie-Division. The December 1944 allocation list shows that ten StuG III were shipped.

PzJg-Abt 208. Attached to 208.Infanterie-Division. The August 1944 allocation list shows that ten StuG III were assigned.

PzJg-Abt 218. Attached to 218.Infanterie-Division. At some time during 1944, possibly September, the battalion received an assault gun company which was referred to as StuG-Abt 1218 and ten StuG III were shipped on 14 December 1944.

PzJg-Abt 225. Attached to 225.Infanterie-Division, the battalion reported five operational StuG IV assault guns on hand with another three in repair. The battalion also reported that one Bergepanzer III recovery tank and two combat-ready StuG III were available, with a further two in transit, but I have been unable to find any record of their allocation (2).

Notes

1. The division was again renamed 45.Volksgrenadier-Division in October 1944.
2. This is yet another example of a unit diary or report which mentions vehicles as being on hand were there is no corresponding entry in the official Zuführungslist, or allocation lists.

A Sturmhaubitze 42 of 3.Batterie, Sturmgeschütz-Abteilung 177 photographed in late spring 1944. Note the extensive concrete reinforcing and Russian T-34 tracks used as extra protection. The battalion was attached to 3.Kavallerie-Brigade at about the same time as this photograph was taken and by August 1944 had been renamed Panzerjäger-Abteilung 69.

PzJg-Abt 227. Attached to 227.Infanterie-Division, the battalion's assault gun company was referred to as StuG-Abt 1227. On 13 December 1944 a total of ten StuG III were sent to the battalion. When the division was withdrawn from the Kurland Pocket in January 1945 all its heavy weapons, including most of the StuG III assault guns, were left behind.

PzJg-Abt 252. Attached to 252.Infanterie-Division. The September 1944 allocation list shows that ten StuG III were shipped in that month.

PzJg-Abt 254. Attached to 254.Infanterie-Division. The August 1944 allocation list shows that ten StuG III were assigned.

PzJg-Abt 291. Attached to 291.Infanterie-Division, the battalion's assault gun company was formed in June 1944 and referred to as StuG-Abt 1291. Ten StuG III were shipped to the battalion during the first week of August 1944 although the assault gun company was reported to be training on the StuG IV at that time.

PzJg-Abt 292. Attached to 292.Infanterie-Division. The September 1944 allocation list shows that ten StuG III were assigned.

PzJg-Abt 248. Attached to 168.Infanterie-Division, the battalion's assault gun company was allocated ten StuG III which were shipped in July 1944.

PzJg-Abt 320. Attached to 320.Infanterie-Division, the battalion was completely destroyed in August 1944 and rebuilt in the following December with three companies. The September 1944 allocation list shows that ten StuG III were shipped in that month.

PzJg-Abt 342. Attached to 342.Infanterie-Division. The September 1944 allocation list shows that ten StuG III were assigned.

PzJg-Abt 371. Attached to 371.Infanterie-Division. The September 1944 allocation list shows that ten StuGIII were assigned.

PzJg-Abt 541. Ten StuG III were allocated to 541.Grenadier-Division, the battalion's parent formation, in July 1944 while it was being formed in Germany.

In October the division was renamed 541.Volksgrenadier-Division and by that time the division's PzJg-Abt had been formed and contained StuG-Abt 1541 and Flak-Kompanie 45. The division served on the Eastern Front from late 1944.

PzJg-Abt 544. The battalion was attached to 544.Grenadier-Division, which was formed in July 1944, and received ten StuG III in the same month. The division served on the Eastern Front and was renamed 544.Volksgrenadier-Division in late October 1944.

The division's PzJg-Abt 1544, renamed PzJg-Abt 544 in February 1945, contained three companies including StuG-Abt 1544.

PzJg-Abt 545. Originally formed as PzJg-Abt 1545 for 545.Grenadier-Division, the battalion was allocated ten StuG III in July 1944. The division served on the Eastern Front and in October 1944 was renamed 545.Volksgrenadier-Division. The battalion's assault gun company was referred to as StuG-Abt 1545.

PzJg-Abt 547. Ten StuG III assault guns were allocated to the battalion's parent formation, 547.Grenadier-Division, in July 1944 but I have been unable to find any further information on this battalion.

PzJg-Abt 548. The battalion was attached to 548.Grenadier-Division, which was formed in July 1944, and received ten StuG III in the same month.

The division served on the Eastern Front and in late October 1944 was renamed 548.Volksgrenadier-Division.

The battalion's assault gun company was referred to as StuG-Abt 1548.

PzJg-Abt 549. Originally formed as PzJg-Abt 1549 for 549.Grenadier-Division, the battalion received ten StuG III in the July 1944. The division was renamed 549.Volksgrenadier-Division in October 1944 and served on the Eastern Front until the end of the war.

In mid February 1945 the battalion was renamed PzJg-Abt 549 and the assault gun company was referred to as StuG-Abt 1549.

PzJg-Abt 551. Originally formed as PzJg-Abt 1551 for 551.Grenadier-Division, the battalion was allocated ten StuG III in August 1944. The division was renamed 551.Volksgrenadier-Division in October 1944 but by this time the battalion had lost all its assault guns (1).

PzJg-Abt 1558. The August 1944 allocation list shows that ten StuG III were allocated to 558.Grenadier-Division and that these vehicles were actually shipped on 1 September. The division was renamed 558.Volksgrenadier-Division in October 1944. Several of these vehicles were still on hand in early 1945 with the battalion's 2.Kompanie.

PzJg-Abt 1561. Originally formed for 561.Grenadier-Division, the battalion was allocated ten StuG III in the September 1944. In early October the division was renamed 561.Volksgrenadier-Division and spent the rest of the war fighting in the East.

PzJg-Abt 1562. Formed in mid-1944 for 562.Grenadier-Division, the battalion was allocated ten StuG III in September of the same year. The division was renamed 562.Volksgrenadier-Division in October 1944 and remained in the East until the end of the war.

Notes

1. The assault guns were eventually replaced with Jagdpanzer 38 tank destroyers.

Below are listed the units of the Waffen-SS which were equipped with assault guns and served on the Eastern Front. I should mention that I have shortened the long and rather complicated titles of these units, some of which were changed several times, as I am sure that most readers will be familiar with these formations.

3.SS-Panzer-Division. This division served exclusively on the Eastern Front and in May 1944 the remnants of the PzJg-Abt, which had been reduced to a single company, were officially incorporated into SS-StuG-Abt 3, although it would seem that the two units had in fact been amalgamated as early as the previous February.

Between late April and early May, StuG-Abt 228 handed over most of its assault guns to the battalion and on 8 June 1944 the division reported that sixteen new assault guns had been received. Just two days later a further three vehicles were handed over from 46.Infanterie-Division and on 18 June another new vehicle arrived. During July a further four assault guns were received and at the same time the battalion was renamed SS-PzJg-Abt 3. In early August the battalion's 1.Kompanie, which had been equipped with towed anti-tank guns, was withdrawn from the front to be retrained and re-equipped with Jagdpanzer IV tank destroyers. The battalion's other two companies continued to operated StuG III assault guns.

From the totals mentioned here, seventeen assault guns were assigned to SS-Pz-Rgt 3 and served with the regiment's 8.Kompanie.

5.SS-Panzer-Division. The division lost all its heavy equipment in the defence and subsequent breakout from the Korsun-Cherkassy Pocket in early 1944 (1).

7.SS-Gebirgs-Division. This division contained a StuG battery made up of seven assault guns from March 1942 until July 1944 when the battery was disbanded, its personnel and equipment incorporated into SS-PzJg-Abt 7 (2). The battalion's assault gun company was referred to as SS-StuG-Abt 1007.

9.SS-Panzer-Division. A full assault gun battalion was planned but it was never completed and the personnel and equipment that had been assembled were merged with II.Abteilung of SS-Pz-Rgt 9 in January 1944. The battalion consisted of two companies of Pzkpfw IV tanks and two companies of StuG III assault guns. When the division returned to the East in early 1945 it reported that sixteen StuG III were on hand.

11.SS-Panzergrenadier-Division. This division was authorised a full assault gun battalion, SS-StuG-Abt 11, made up of three companies in late 1943 and forty-two StuG III had been dispatched by the end of the year (3). At some time in 1944 the battalion was renamed SS-PzJg-Abt 11 and this unit did not return to its parent formation until February 1945. It is very likely that this battalion received the thirty-one StuG III shipped in the previous month. In late 1944 the division's SS-Pz-Abt 11 was converted to a Pz-StuG-Abt and by January 1945 was equipped with a mixed bag of Pzkpfw V Panther ausf D (4) tanks and eight assault guns concentrated in a single company.

Notes

1. Contrary to many accounts this division did not operate any Sturmgeschütz III assault guns during 1944 other than those lost in the Korsun-Cherkassy battles.
2. According to the post-war account of Brigadeführer Otto Kumm, a former commander of the division, the battalion still had at least twelve assault guns on hand in October 1944 when it was fighting south of Belgrade.
3. A handwritten amendment in the October 1943 allocation list shows that fourteen StuG III intended for this division were diverted to 1.SS-Panzer-Division.
4. The planned Pzkpfw IV delivery never materialised and the tanks mentioned here were taken over from Panzer-Regiment 29 which had inherited these elderly models from SS-Panzer-Regiment 1.

Photographed in Yugoslavia during the autumn of 1944, this Sturmgeschütz III ausf G is from SS-Panzerjäger-Abteilung 7. The method of holding the spare road wheels and the large stowage box are both distinctive features of the battalion's assault guns. Note the enlarged, non-standard aperture of the MG 34 shield and the metal tabs at the front and rear of the superstructure that held the Schürzen rails.

Notes

1. These assault guns may in fact have been vehicles of SS-StuG-Batterie Nord.
2. Photographic evidence would suggest that a number of these vehicles were early models equipped with the short-barelled L/24 gun.
3. These formations remained under the authority of the Luftwaffe, which is to say Reichsmarschall Göring, and maintained their uniform distinctions and rank structure.
4. These two divisions, with several support units, formed PanzerKorps HG and the information given here is a greatly simplified version of their complex histories. The prefix Fallschirm was used to signify that they were very much a part of the Luftwaffe but should not be taken mean that they were parachute units.

18.SS-Panzergrenadier-Division. By early July 1944, when this division participated in the occupation of Hungary, it was able to field a strong battle group only, made up of armoured infantry and reconnaissance units with a single company of ten StuG III assault guns controlled by SS-Pz-Abt 18 (1). In August and September 1944 a total of twenty-six Stug III were shipped to the division and it was apparently intended that these vehicles would equip SS-PzJg-Abt 18 but they were all eventually assigned to the Panzer battalion (2). By December just six assault guns were on hand and in the same month the division was withdrawn from the front.

SS-PzJg-Abt 54. Raised mainly from Dutch volunteers in late October 1943, the battalion remained semi-autonomous throughout most of 1944 although it had been intended that it would support 4.SS-Panzergrenadier-Brigade. In early 1945 it was renamed and permanently attached to 23.SS-Panzergrenadier-Divison.

In late 1942 it was suggested that surplus personnel from the Luftwaffe be used to replace the large number of casualties suffered on the Russian Front and whole divisions, referred to as Luftwaffen-Feld-Divisionen, were raised (3). During the time period covered by our narrative most were serving in the West and are not included here.

The only significant armoured units under Luftwaffe control were the two Hermann Göring (HG) divisions.

Fallschirm-Panzer-Division 1 HG. The division was formed in Italy in July 1944 from units of Panzer-Division HG and was very soon afterwards transferred from Italy to the East. The division's assault guns were operated by Fallschirm-StuG-Abt HG which reported forty-five StuG III on hand in mid-October 1944. On 1 January 1945, when the division was fighting in the Weichsel bend as part of Hgr Mitte, the battalion was able to field thirty-two assault guns.

Fallschirm-Panzergrenadier-Division 2 HG. On 1 December 1944 the division's commander reported that twenty-nine operational StuG III were available with Fallschirm-StuG-Abt 2 HG (4).

In April 1944 Luftwaffen-Feld-Divisionen were renamed Feld-Divisionen (L) des H. Note that only those units which served in the East are mentioned here.

Feld-Division (L).12. In July 1944 the division's PzJg-Abt 12 (L) reported that eight StuG III were available. The battalion's assault gun company was referred to as StuG-Ab 2012(L) and was reinforced by elements of Feld-Division (L).13 which had been disbanded in April 1944.

In late 1944 and early 1945 the division was fighting in Kurland and ended the war in the Danzig-Gotenhafen area.

Feld-Division (L).21. In May 1944 the division's PzJg-Abt 21 (L) reported that eight StuG III assault guns were on hand with the second company. A company of Jagdpanzer 38(t) tank destroyers originally allocated to the division as Jagdpanzer-Kompanie 1021(L) never materialised and was instead appropriated by Panzer-Jagd-Abteilung 1, an independent unit attached to Hgr Weichsel.

By January 1945 the battalion reported that ten assault guns were available for operations.

Photographed outside Kiev in early 1944, after the fighting for the Korsun–Cherkassy Pocket, this MIAG-built Sturmgeschütz III of Sturmgeschütz-Abteilung 202 provides a good example of how early production vehicles remained in service for some time. Note the first pattern muzzle brake, the drive sprocket cover, the lack of an armoured deflector for the commander's cupola and the complete absence of Zimmerit. The bracket holding the lengths of spare track and the rack on the rear hull were modifications introduced by this battalion. The name Tigerhai translates as Tiger Shark.

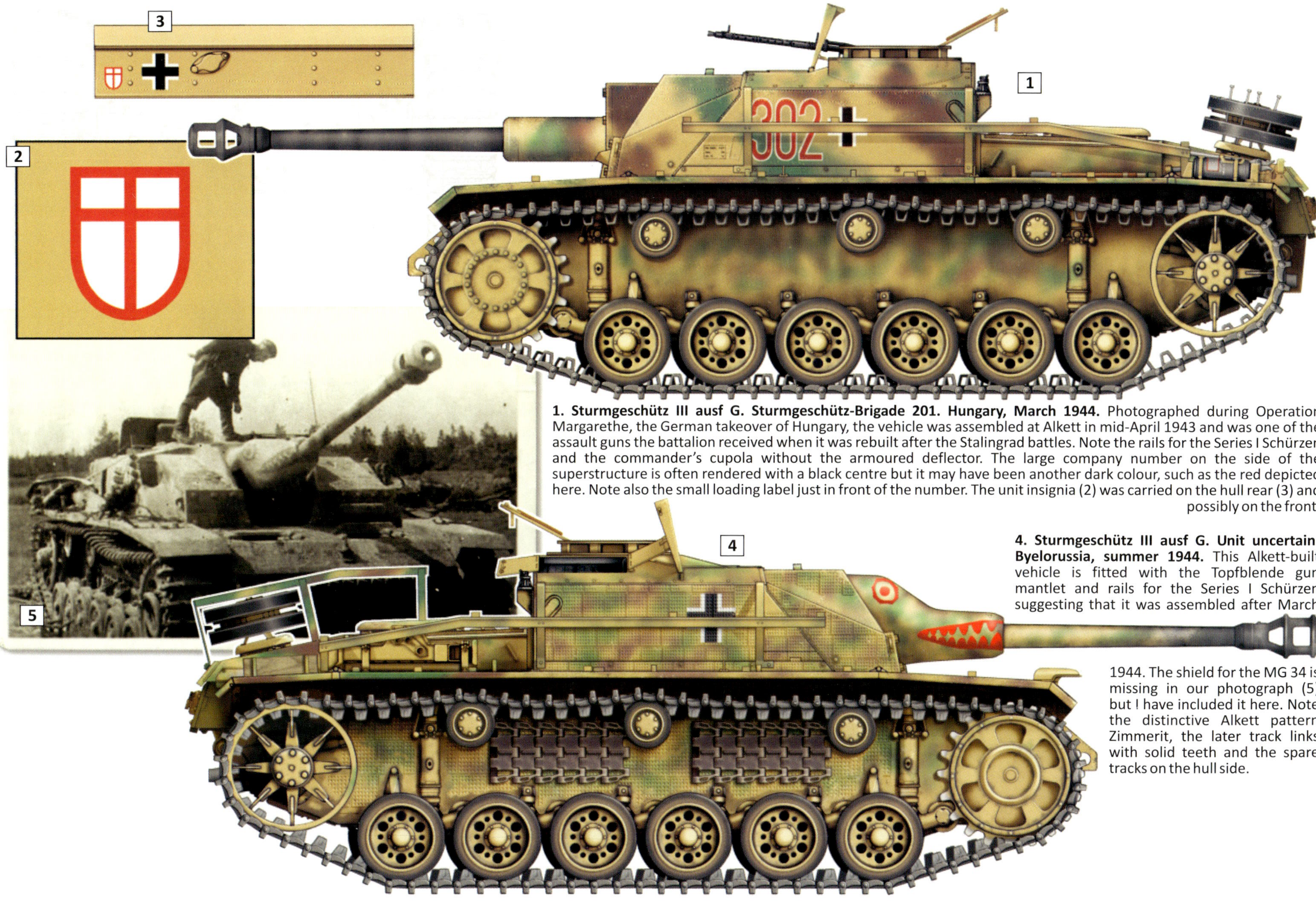

1. Sturmgeschütz III ausf G. Sturmgeschütz-Brigade 201. Hungary, March 1944. Photographed during Operation Margarethe, the German takeover of Hungary, the vehicle was assembled at Alkett in mid-April 1943 and was one of the assault guns the battalion received when it was rebuilt after the Stalingrad battles. Note the rails for the Series I Schürzen and the commander's cupola without the armoured deflector. The large company number on the side of the superstructure is often rendered with a black centre but it may have been another dark colour, such as the red depicted here. Note also the small loading label just in front of the number. The unit insignia (2) was carried on the hull rear (3) and possibly on the front.

4. Sturmgeschütz III ausf G. Unit uncertain. Byelorussia, summer 1944. This Alkett-built vehicle is fitted with the Topfblende gun mantlet and rails for the Series I Schürzen suggesting that it was assembled after March 1944. The shield for the MG 34 is missing in our photograph (5) but I have included it here. Note the distinctive Alkett pattern Zimmerit, the later track links with solid teeth and the spare tracks on the hull side.

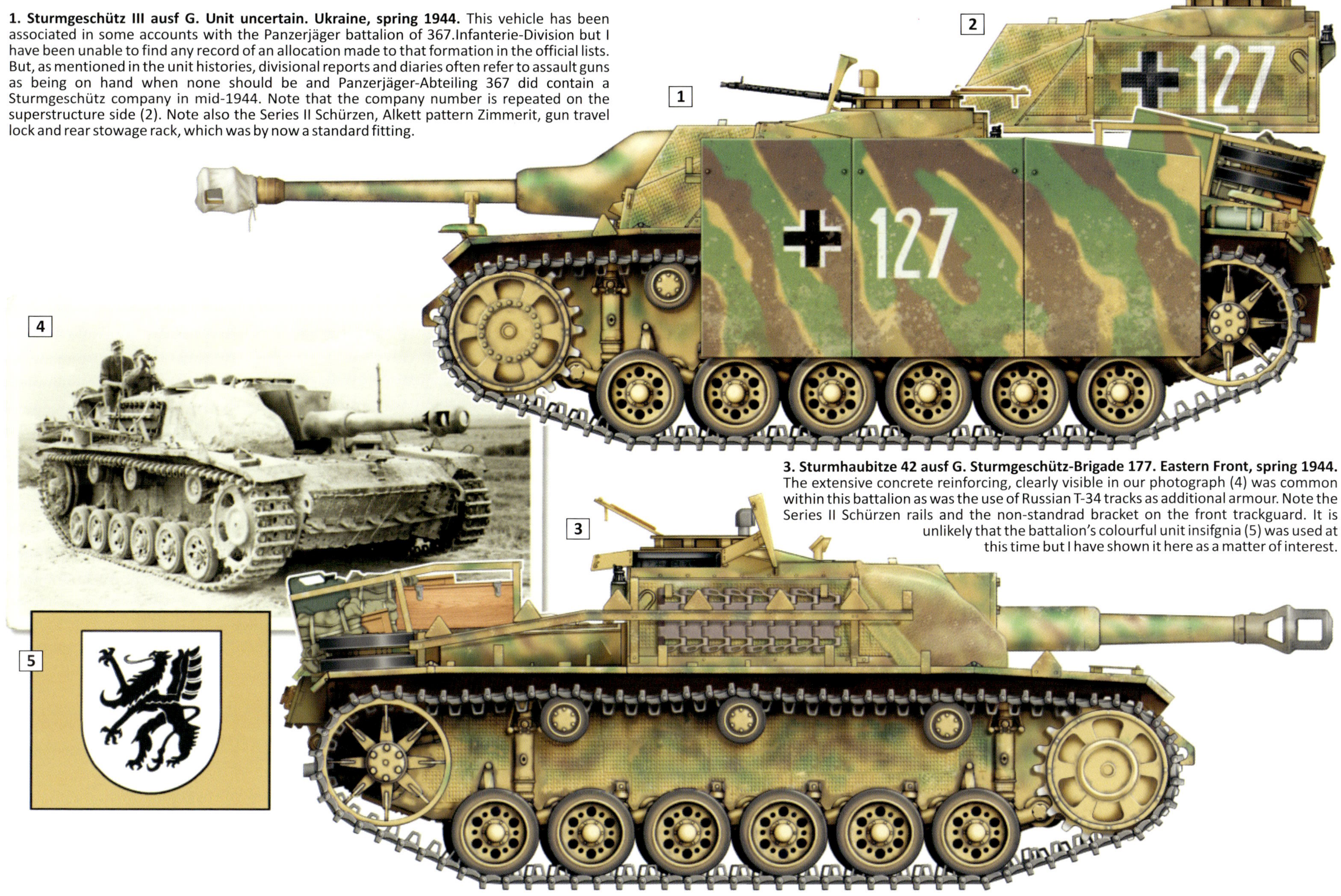

1. Sturmgeschütz III ausf G. Unit uncertain. Ukraine, spring 1944. This vehicle has been associated in some accounts with the Panzerjäger battalion of 367.Infanterie-Division but I have been unable to find any record of an allocation made to that formation in the official lists. But, as mentioned in the unit histories, divisional reports and diaries often refer to assault guns as being on hand when none should be and Panzerjäger-Abteiling 367 did contain a Sturmgeschütz company in mid-1944. Note that the company number is repeated on the superstructure side (2). Note also the Series II Schürzen, Alkett pattern Zimmerit, gun travel lock and rear stowage rack, which was by now a standard fitting.

3. Sturmhaubitze 42 ausf G. Sturmgeschütz-Brigade 177. Eastern Front, spring 1944. The extensive concrete reinforcing, clearly visible in our photograph (4) was common within this battalion as was the use of Russian T-34 tracks as additional armour. Note the Series II Schürzen rails and the non-standrad bracket on the front trackguard. It is unlikely that the battalion's colourful unit insifgnia (5) was used at this time but I have shown it here as a matter of interest.

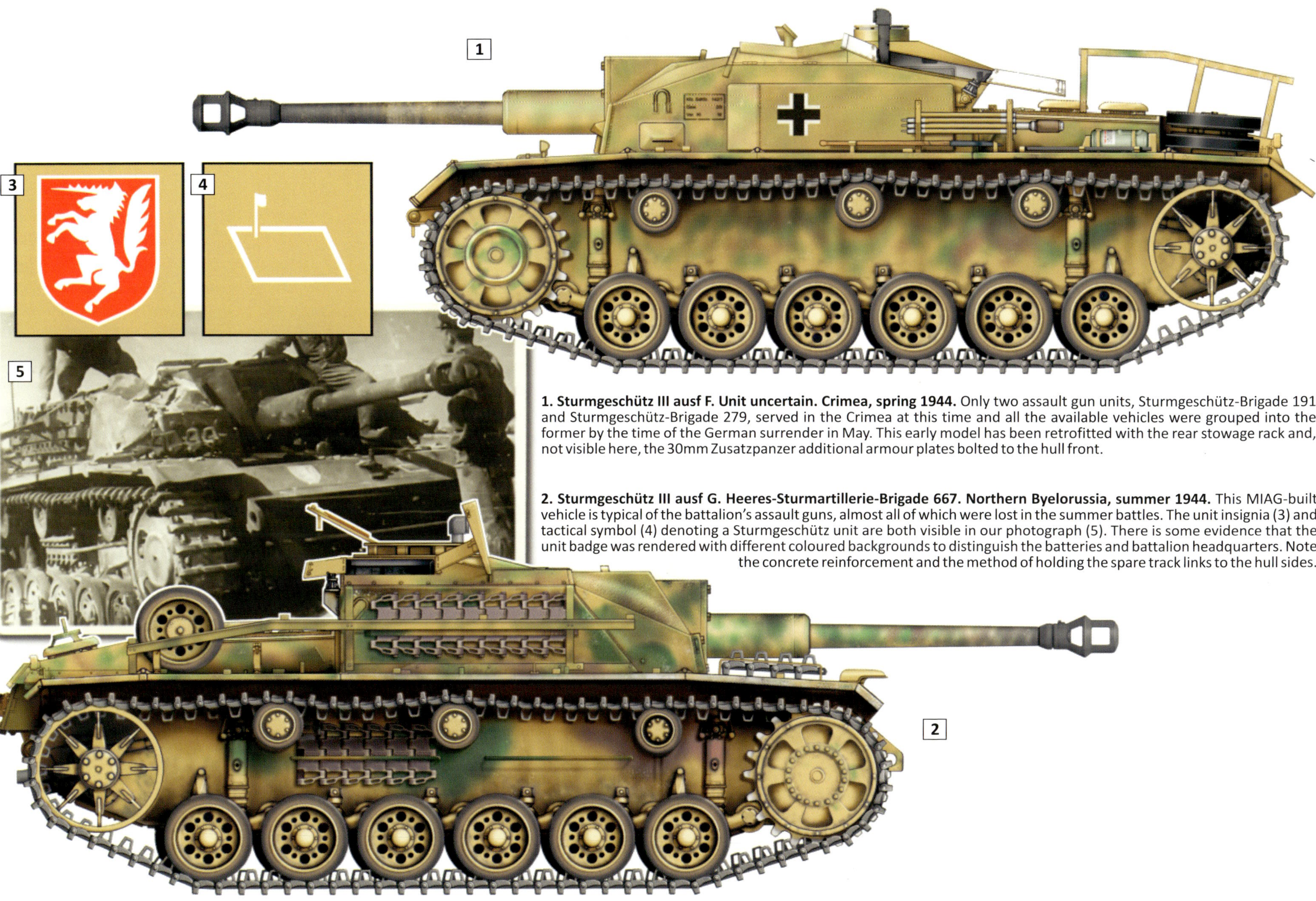

1. Sturmgeschütz III ausf F. Unit uncertain. Crimea, spring 1944. Only two assault gun units, Sturmgeschütz-Brigade 191 and Sturmgeschütz-Brigade 279, served in the Crimea at this time and all the available vehicles were grouped into the former by the time of the German surrender in May. This early model has been retrofitted with the rear stowage rack and, not visible here, the 30mm Zusatzpanzer additional armour plates bolted to the hull front.

2. Sturmgeschütz III ausf G. Heeres-Sturmartillerie-Brigade 667. Northern Byelorussia, summer 1944. This MIAG-built vehicle is typical of the battalion's assault guns, almost all of which were lost in the summer battles. The unit insignia (3) and tactical symbol (4) denoting a Sturmgeschütz unit are both visible in our photograph (5). There is some evidence that the unit badge was rendered with different coloured backgrounds to distinguish the batteries and battalion headquarters. Note the concrete reinforcement and the method of holding the spare track links to the hull sides.

1

2

1. Sturmgeschütz III ausf G. SS-Panzerjäger-Abteilung 54. Estonia, summer 1944. Commanded by Oberscharführer Johannes Cuyper, a Dutch volunteer, this MIAG-built assault gun is probably one of the first the battalion received in late 1943. Note that the Series I Schürzen does not have the additional centre armour plates, although they may be fixed on the inside. The battalion was permanently attached to 23.SS-Freiwilligen-Panzergrenadier-Division Nederland and the division's unit insignia is shown here (2).

3. Sturmgeschütz III ausf G. Panzerjäger-Abteilung 28. Northern Ukraine, summer 1944. This MIAG-Built assault gun may be one of the original allocation of fourteen vehicles made to the battalion's parent formation, 28.Jäger-Division, in July 1943. Note the Series I Schürzen and the very rough application of Zimmerit which is particularly noticeable in the front view (4). Unit insignia (5) incorporating an abbreviated version of the formation's designation with the appropriate tactical symbol became increasingly common from this time, especially in the East.

3

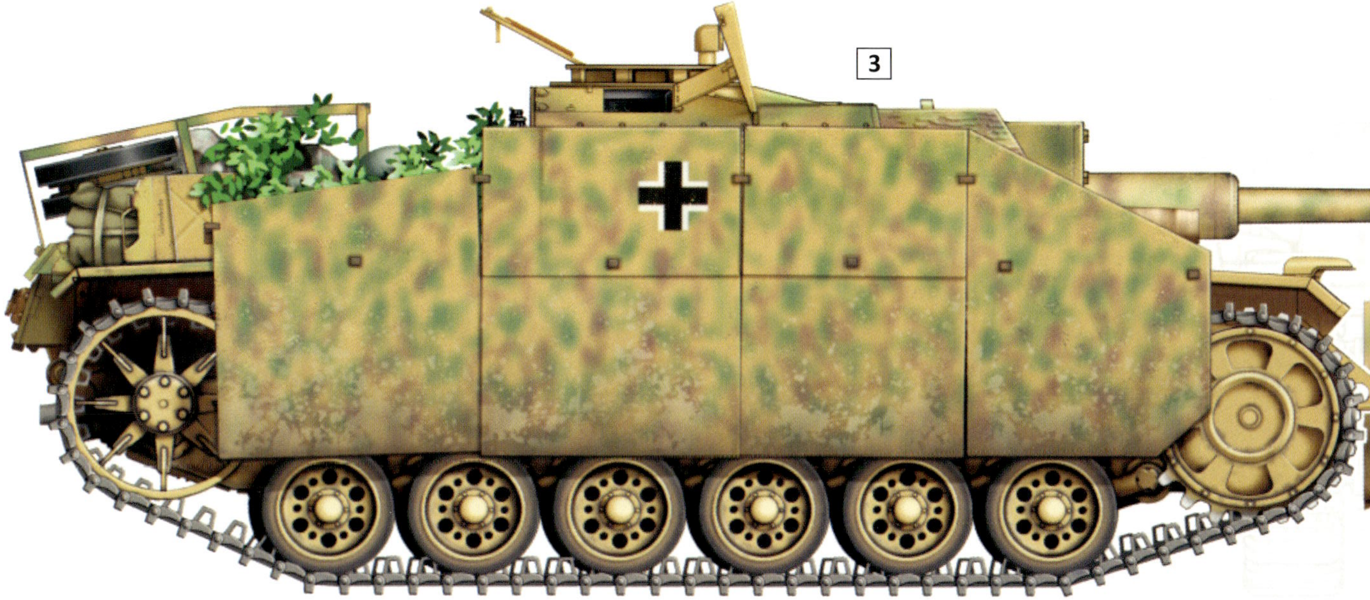

4

5

St.Gesch.Abt. 1028

1. Sturmgeschütz III ausf G. Sturmgeschütz-Brigade 303. Finland, summer 1944. Photographed during the fighting for Ihantala in southern Finland in June 1944, this assault gun is fitted with the cast steel return rollers introduced into production in November 1943. Note the rack for spare tracks bolted to the hull side and the logs, a common feature on this sector of the front. The name painted on the Topfblende mantlet is unfortunately indecipherable but is just visible in our photograph (2). The second vehicle (3) in the photograph on which this illustration is based has a crudely applied number 44 on the rear hull and it is tempting to speculate that this may denote the proposed fourth battery which most sources state was never established .

4. Sturmgeschütz III ausf G. Sturmgeschütz-Brigade 303. Finland, summer 1944. This vehicle is also shown in the photograph on page 8 and discussed further there. Of note is the protective armoured hood (5) over the driver's visor.

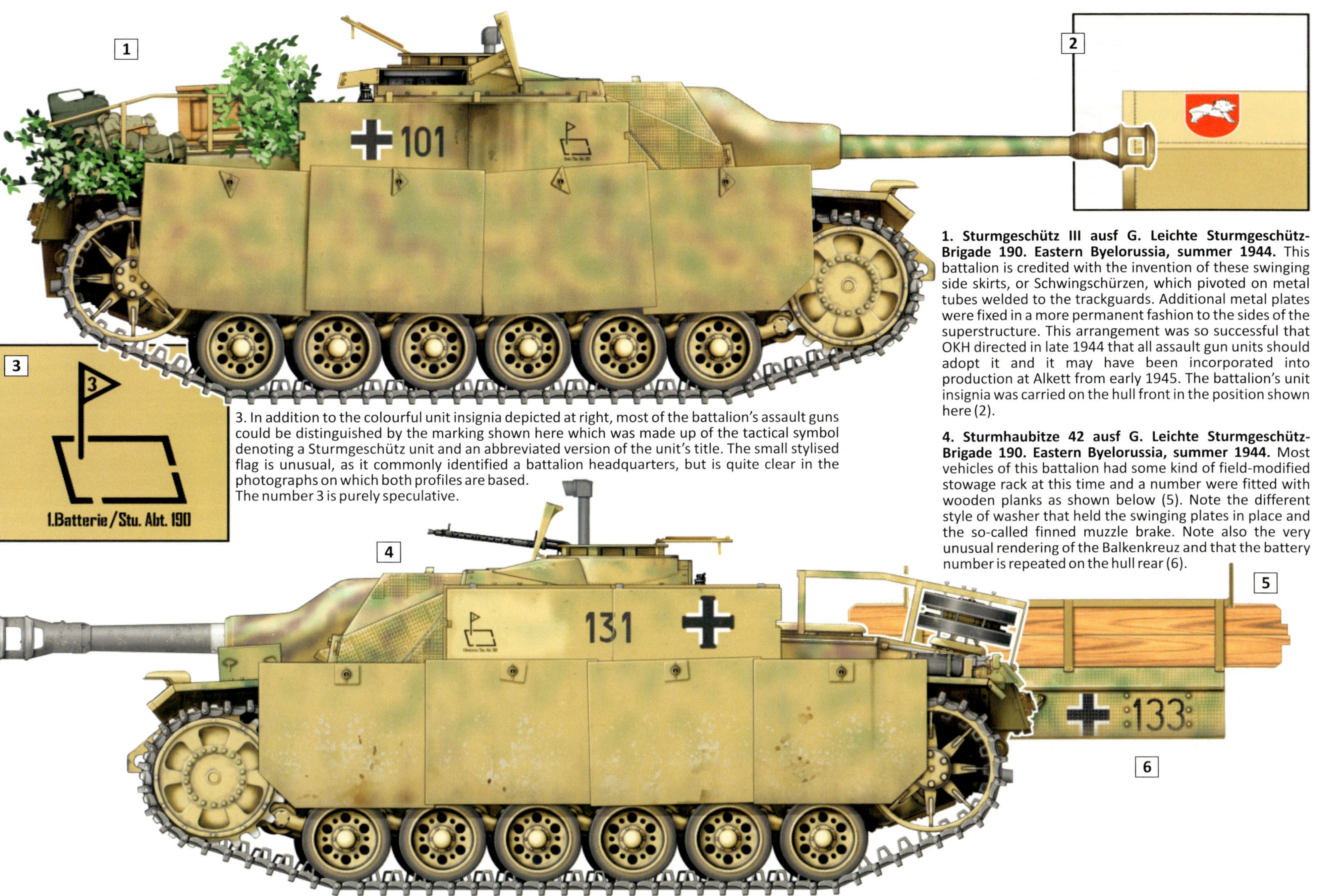

1. Sturmgeschütz III ausf G. Leichte Sturmgeschütz-Brigade 190. Eastern Byelorussia, summer 1944. This battalion is credited with the invention of these swinging side skirts, or Schwingschürzen, which pivoted on metal tubes welded to the trackguards. Additional metal plates were fixed in a more permanent fashion to the sides of the superstructure. This arrangement was so successful that OKH directed in late 1944 that all assault gun units should adopt it and it may have been incorporated into production at Alkett from early 1945. The battalion's unit insignia was carried on the hull front in the position shown here (2).

3. In addition to the colourful unit insignia depicted at right, most of the battalion's assault guns could be distinguished by the marking shown here which was made up of the tactical symbol denoting a Sturmgeschütz unit and an abbreviated version of the unit's title. The small stylised flag is unusual, as it commonly identified a battalion headquarters, but is quite clear in the photographs on which both profiles are based.
The number 3 is purely speculative.

4. Sturmhaubitze 42 ausf G. Leichte Sturmgeschütz-Brigade 190. Eastern Byelorussia, summer 1944. Most vehicles of this battalion had some kind of field-modified stowage rack at this time and a number were fitted with wooden planks as shown below (5). Note the different style of washer that held the swinging plates in place and the so-called finned muzzle brake. Note also the very unusual rendering of the Balkenkreuz and that the battery number is repeated on the hull rear (6).

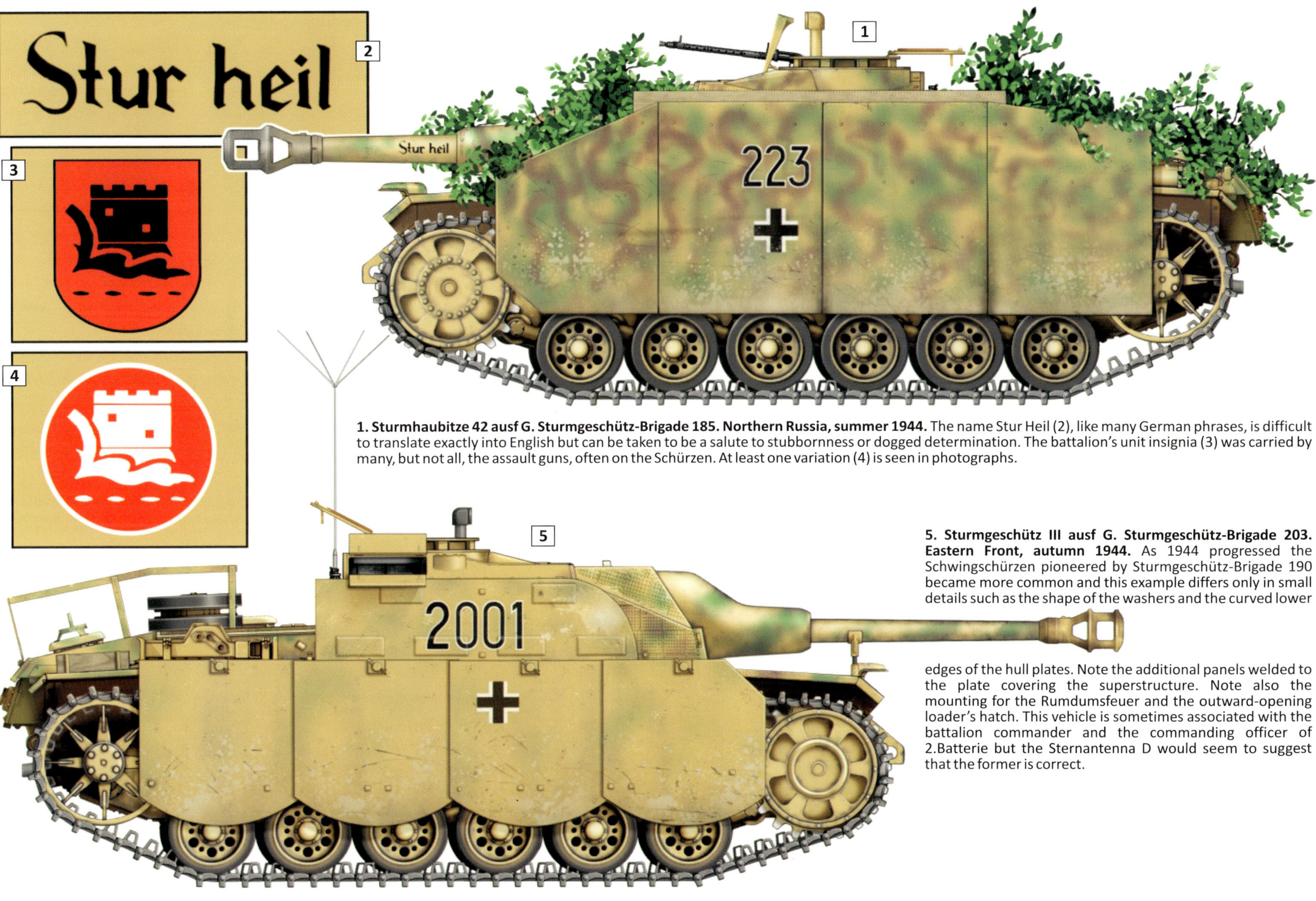

1. Sturmhaubitze 42 ausf G. Sturmgeschütz-Brigade 185. Northern Russia, summer 1944. The name Stur Heil (2), like many German phrases, is difficult to translate exactly into English but can be taken to be a salute to stubbornness or dogged determination. The battalion's unit insignia (3) was carried by many, but not all, the assault guns, often on the Schürzen. At least one variation (4) is seen in photographs.

5. Sturmgeschütz III ausf G. Sturmgeschütz-Brigade 203. Eastern Front, autumn 1944. As 1944 progressed the Schwingschürzen pioneered by Sturmgeschütz-Brigade 190 became more common and this example differs only in small details such as the shape of the washers and the curved lower edges of the hull plates. Note the additional panels welded to the plate covering the superstructure. Note also the mounting for the Rumdumsfeuer and the outward-opening loader's hatch. This vehicle is sometimes associated with the battalion commander and the commanding officer of 2.Batterie but the Sternantenna D would seem to suggest that the former is correct.

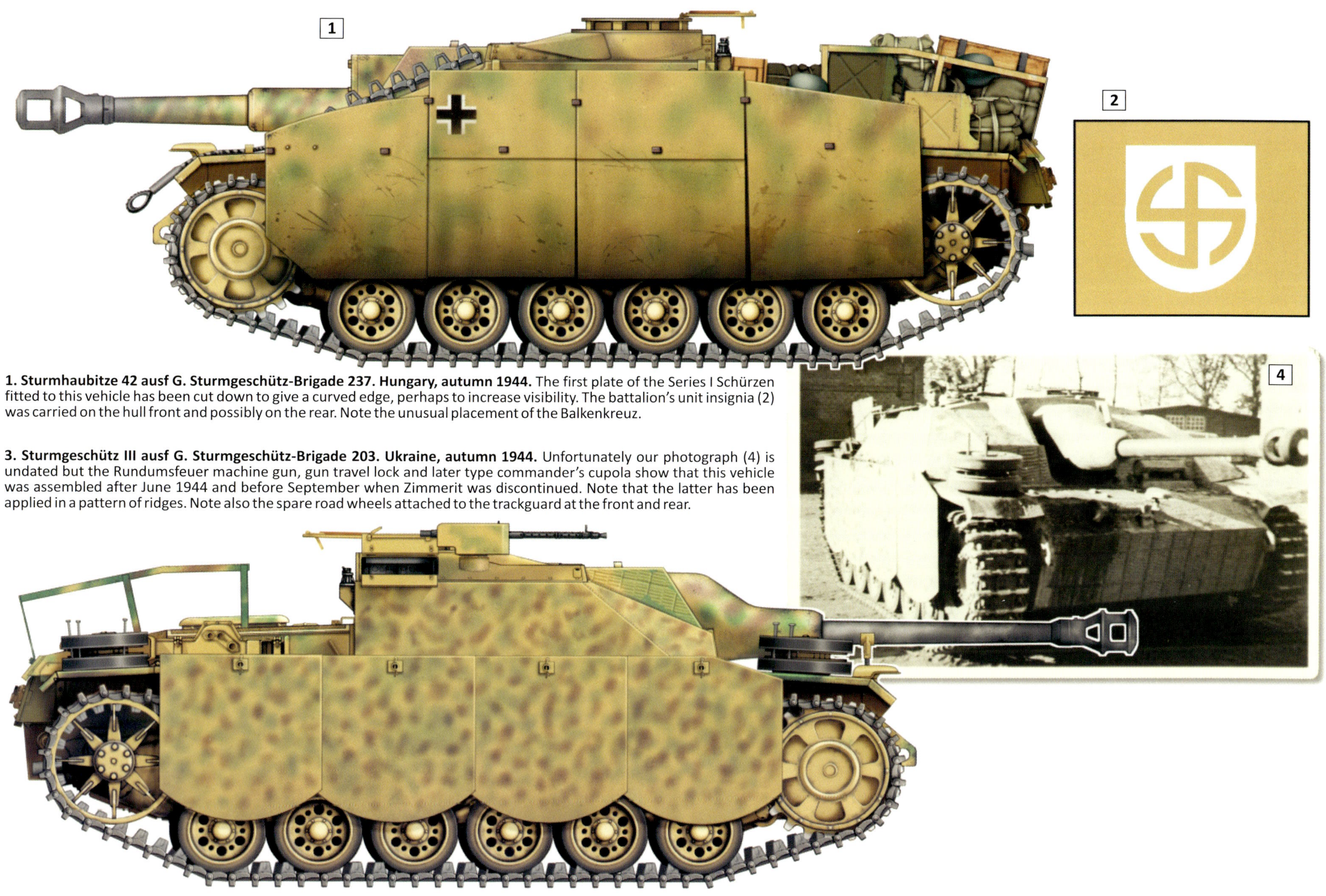

1. Sturmhaubitze 42 ausf G. Sturmgeschütz-Brigade 237. Hungary, autumn 1944. The first plate of the Series I Schürzen fitted to this vehicle has been cut down to give a curved edge, perhaps to increase visibility. The battalion's unit insignia (2) was carried on the hull front and possibly on the rear. Note the unusual placement of the Balkenkreuz.

3. Sturmgeschütz III ausf G. Sturmgeschütz-Brigade 203. Ukraine, autumn 1944. Unfortunately our photograph (4) is undated but the Rundumsfeuer machine gun, gun travel lock and later type commander's cupola show that this vehicle was assembled after June 1944 and before September when Zimmerit was discontinued. Note that the latter has been applied in a pattern of ridges. Note also the spare road wheels attached to the trackguard at the front and rear.

1. Sturmgeschütz III ausf F/8. Sturmgeschütz-Ersatz-Abteilung 200. Poland, autumn 1944. This unit was little more than a company and according to some sources could muster just six assault guns when it took part in the fighting in Warsaw. Many of the factory fittings have been removed and the battery number has been painted over an older version as can be seen in our photograph (2)

3. Sturmgeschütz III ausf G. Unit uncertain. Poland, autumn 1944. This vehicle is fitted with the earliest version of the Series I Schürzen identified by the straight lower edge of the first plate. Although it is difficult to be certain, this vehicle may be one of the assault guns assigned to either SS-Panzerjäger-Abteilung 3 or 8.Kompanie, SS-Panzer-Regiment 3. The division's well known unit insignia (4) was carried in the position shown here (5) in late 1943 and may have been retained into the following year.

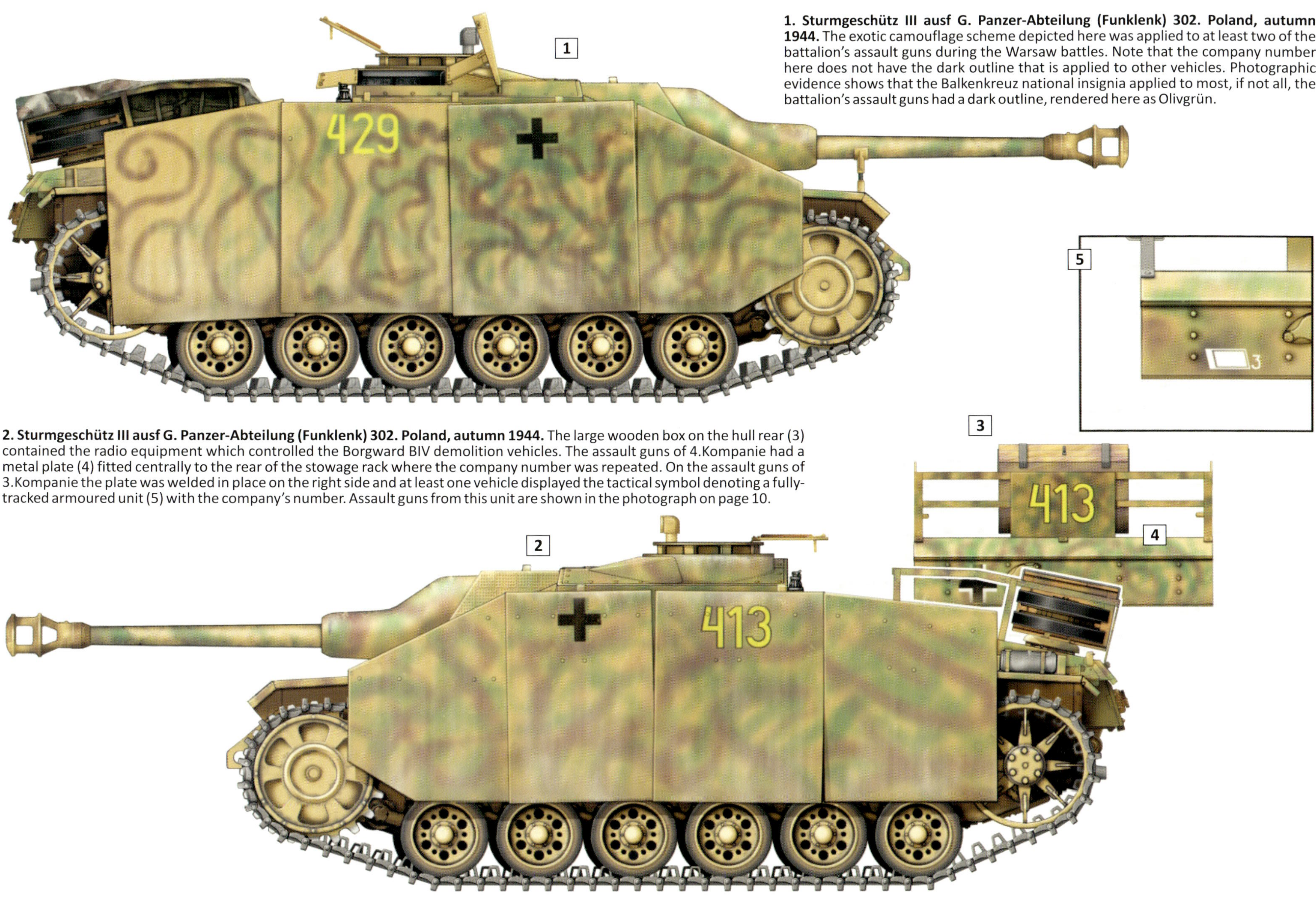

1. Sturmgeschütz III ausf G. Panzer-Abteilung (Funklenk) 302. Poland, autumn 1944. The exotic camouflage scheme depicted here was applied to at least two of the battalion's assault guns during the Warsaw battles. Note that the company number here does not have the dark outline that is applied to other vehicles. Photographic evidence shows that the Balkenkreuz national insignia applied to most, if not all, the battalion's assault guns had a dark outline, rendered here as Olivgrün.

2. Sturmgeschütz III ausf G. Panzer-Abteilung (Funklenk) 302. Poland, autumn 1944. The large wooden box on the hull rear (3) contained the radio equipment which controlled the Borgward BIV demolition vehicles. The assault guns of 4.Kompanie had a metal plate (4) fitted centrally to the rear of the stowage rack where the company number was repeated. On the assault guns of 3.Kompanie the plate was welded in place on the right side and at least one vehicle displayed the tactical symbol denoting a fully-tracked armoured unit (5) with the company's number. Assault guns from this unit are shown in the photograph on page 10.

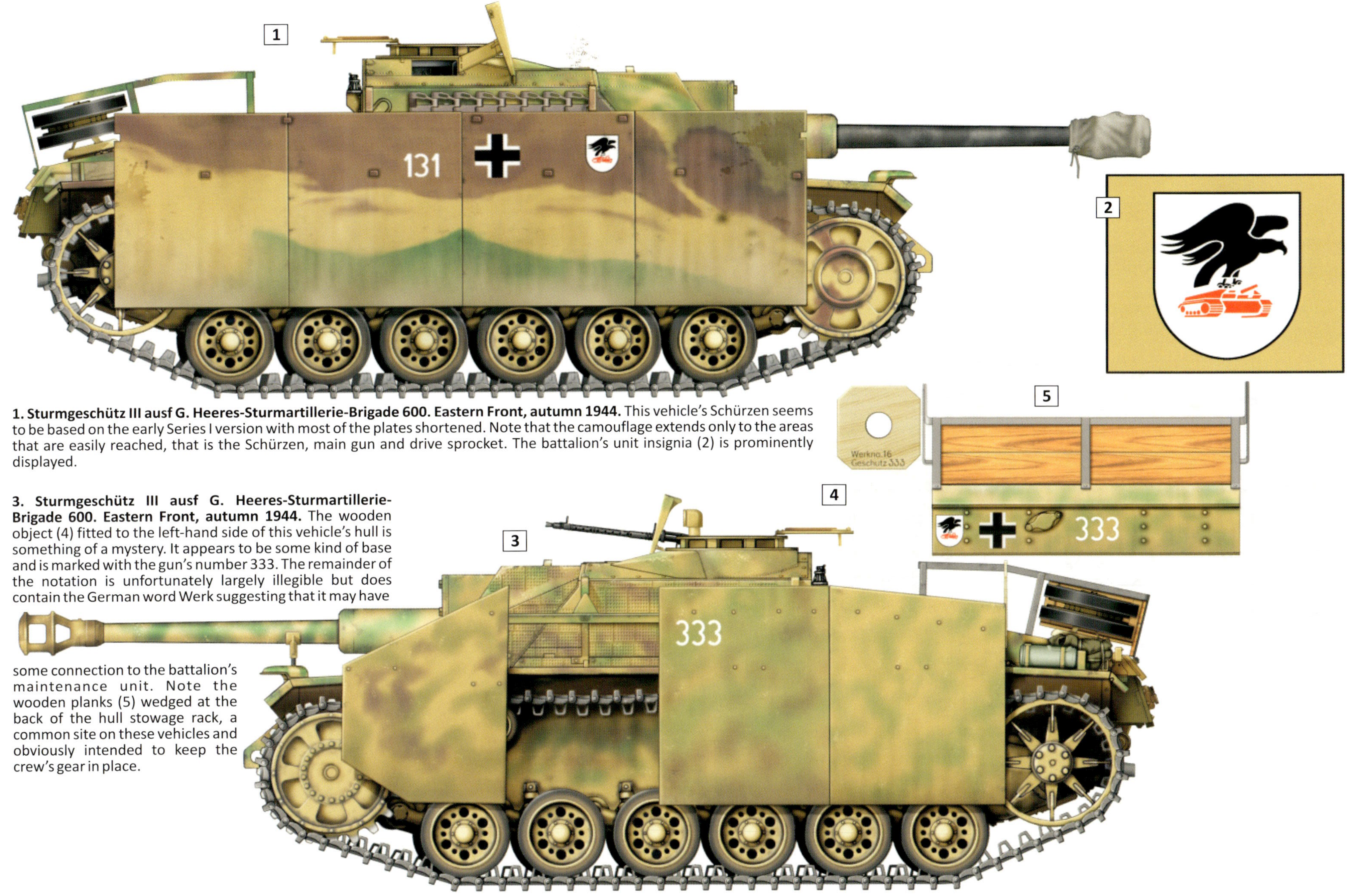

1. Sturmgeschütz III ausf G. Heeres-Sturmartillerie-Brigade 600. Eastern Front, autumn 1944. This vehicle's Schürzen seems to be based on the early Series I version with most of the plates shortened. Note that the camouflage extends only to the areas that are easily reached, that is the Schürzen, main gun and drive sprocket. The battalion's unit insignia (2) is prominently displayed.

3. Sturmgeschütz III ausf G. Heeres-Sturmartillerie-Brigade 600. Eastern Front, autumn 1944. The wooden object (4) fitted to the left-hand side of this vehicle's hull is something of a mystery. It appears to be some kind of base and is marked with the gun's number 333. The remainder of the notation is unfortunately largely illegible but does contain the German word Werk suggesting that it may have some connection to the battalion's maintenance unit. Note the wooden planks (5) wedged at the back of the hull stowage rack, a common site on these vehicles and obviously intended to keep the crew's gear in place.

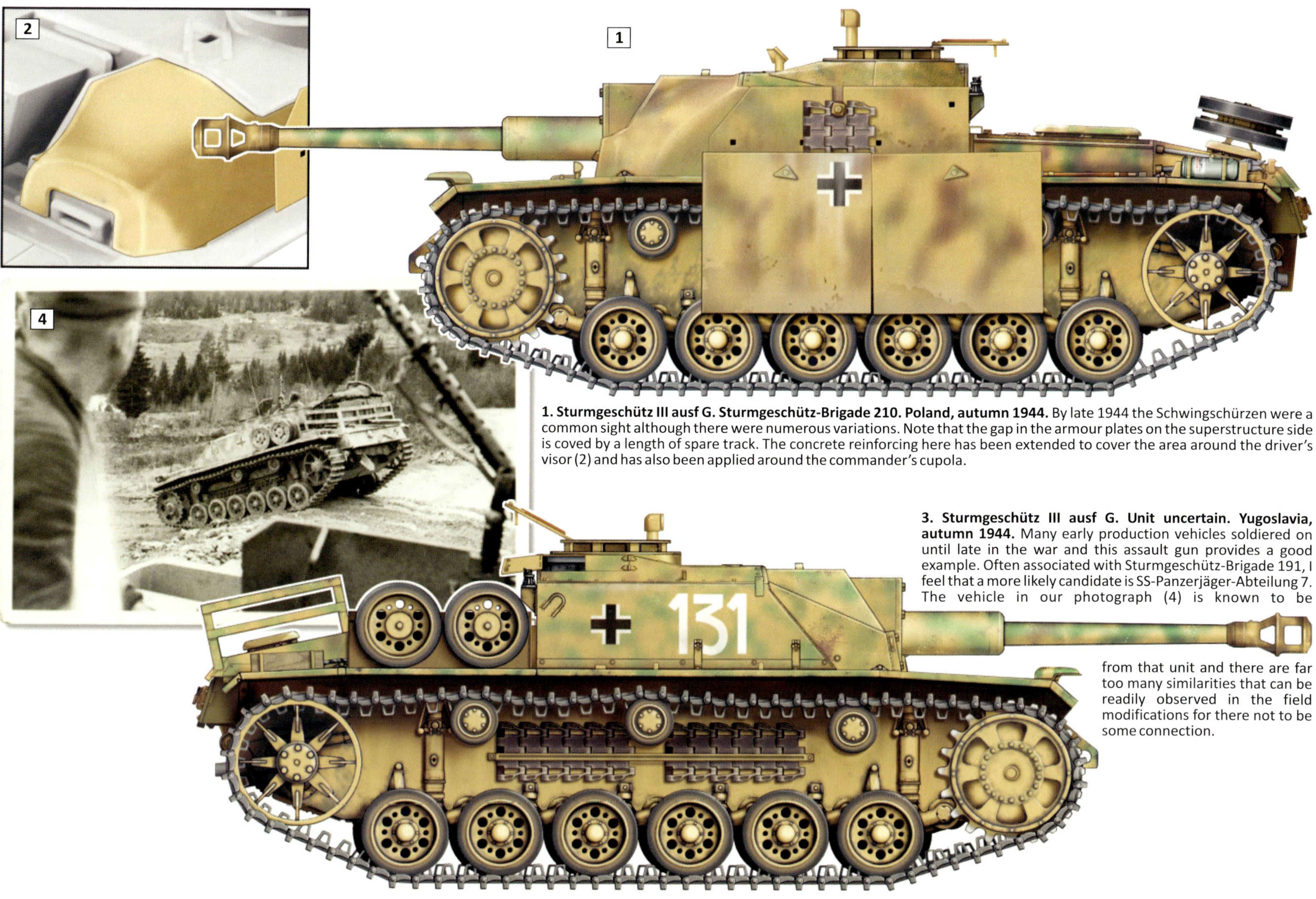

1. Sturmgeschütz III ausf G. Sturmgeschütz-Brigade 210. Poland, autumn 1944. By late 1944 the Schwingschürzen were a common sight although there were numerous variations. Note that the gap in the armour plates on the superstructure side is coved by a length of spare track. The concrete reinforcing here has been extended to cover the area around the driver's visor (2) and has also been applied around the commander's cupola.

3. Sturmgeschütz III ausf G. Unit uncertain. Yugoslavia, autumn 1944. Many early production vehicles soldiered on until late in the war and this assault gun provides a good example. Often associated with Sturmgeschütz-Brigade 191, I feel that a more likely candidate is SS-Panzerjäger-Abteilung 7. The vehicle in our photograph (4) is known to be from that unit and there are far too many similarities that can be readily observed in the field modifications for there not to be some connection.

STUG III

KEVIN D POTTS

1/35 SCALE

Regular readers of this series will be familiar with Kevin's highly detailed and evocative models. For this project he used the Dragon Models 1/35 scale StuG III ausf G w/Zimmerit, shown at left.

Below: The finished product, less the commander figure, ready for painting and weathering. Any gaps in the final assembly were filled with stretched plastic sprue, a method Kevin found to be more effective than the use of putty in preserving the texture of the Zimmerit. The metal storage box on the engine deck was made from styrene card with the addition of photo-etched brass hinges and a number of the other parts were taken from Kevin's spares box.

Below, left to right: The reworked crank starter port in the open position. Interior details including the vehicle's radio equipment, MP 40 and ventilator. These are all visible through the open loader's hatch.

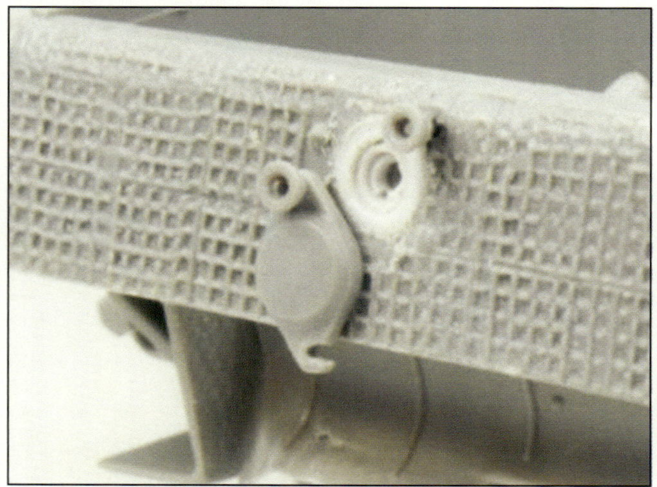

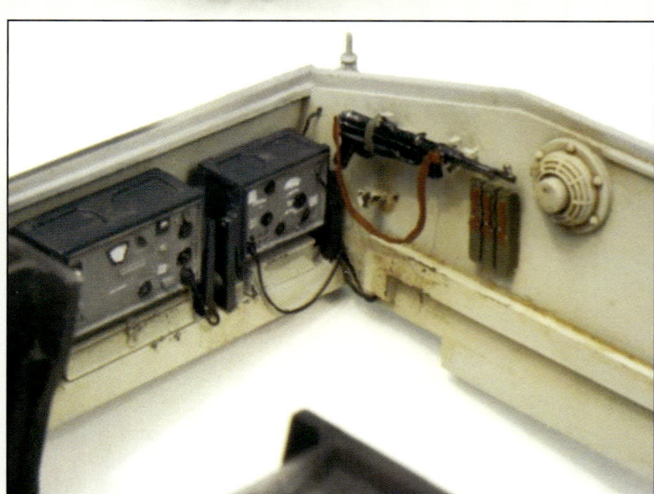

A front view of Kevin's completed model with details below showing: 1. The Notek headlight and gun travel lock 2. The Rundumsfeuer machine gun and its shield. 3. The inner face of the series II Schürzen plates with their supporting rails. 4. Spare road wheel, partly held in place with track pins. 5. The vehicle's fire extinguisher. 6. Part of the rear stowage racks and the protective metal grills of the air-intake system.

Kevin's fully-assembled and painted StuG III model. It was first treated with a grey primer coat and black was added in the areas which would be normally in shade. Once thoroughly dry, this was followed with Model Master Dark Yellow and Vallejo Model Color Green and Brown. The small dots that complete the ambush pattern camouflage scheme were painted by hand with a fine brush.

Artists oil paints and AK Interactive enamels were used for weathering and to pick out the panel lines. Kevin mixed acrylic paste and tinted it with Vallejo Model Color paints to replicate dirt and mud. The commander figure is an Alpine Miniatures body with a Hornet Models head with wire added for the headset. The stencilling on the ammunition crate is from Archer Fine Transfers.

STUH 42

JOSEPH YOUNGERMAN

1/48 SCALE

At left: Joseph's replica is based on the Tamiya 1/48 scale Sturmgeschütz III marketed as the Frühe, or early, version. Below: Built straight from the box, the model was fitted with the 10.5cm gun as both this weapon and parts for the 7.5cm L/48 are included in the kit. The partially completed model was painted in dark yellow and Joseph used the transfers supplied with the kit to finish the markings. Note that a number of early production features, such as the smoke grenade dischargers were deliberately omitted.

Above: Details of the partly-assembled and painted model including the engine deck, spare wheel and hull front. Note the fine application of mud on the front fender. Below: Joseph's StuH 42 almost finished with the top and bottom halves joined.

Both images here show the fully completed model. A number of washes were added to achieve the worn and slightly damaged effect including Tamiya Panel Line Accent Color and other weathering products from AK Interactive and AMMO by Mig were also employed. Smaller details were rendered with a fine brush. This is a particularly impressive effort as it was Joseph's first armour model.

STUG III

MASAHIRO DOI

1/30 SCALE

Masahiro's model is based on the Nichimo 1/30 scale Pzkpfw III ausf M kit which was first released in 1971. Some elements of the Nichimo German Infantry set were used to build the crew figures. The superstructure and all its fittings were created from plastic card, brass rod and metal sheet.

Below: Some examples of the incredible details that were scratchbuilt by Masahiro including the commander's cupola, loader's hatch and MG shield, Notek headlight, trackguard support, bolted-on armour, exhaust pipe and rear towing coupling.

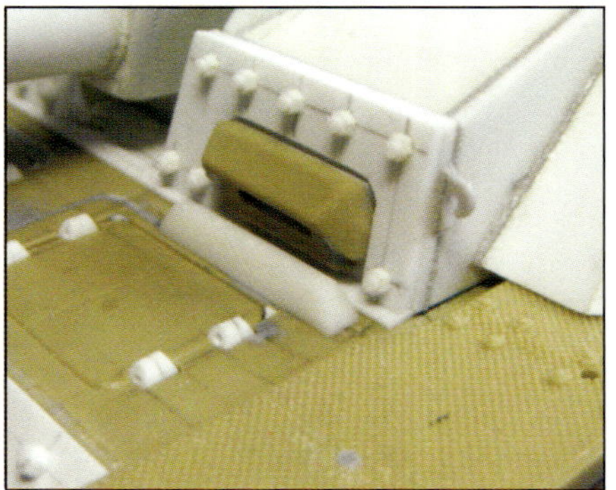
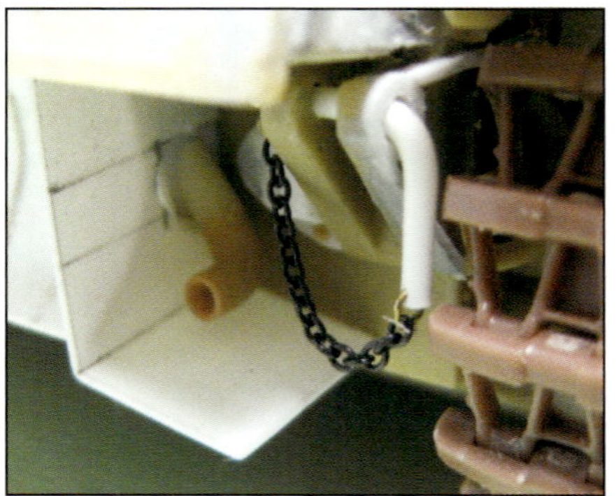

Above: The complete assembly ready for painting. The Schürzen are detachable and were painted separately. Below: Details of the hull fittings including the Schürzen rails and their support brackets, the left side air intake with its protective grill and the right side air intake and scratchbuilt jack and starter crank.

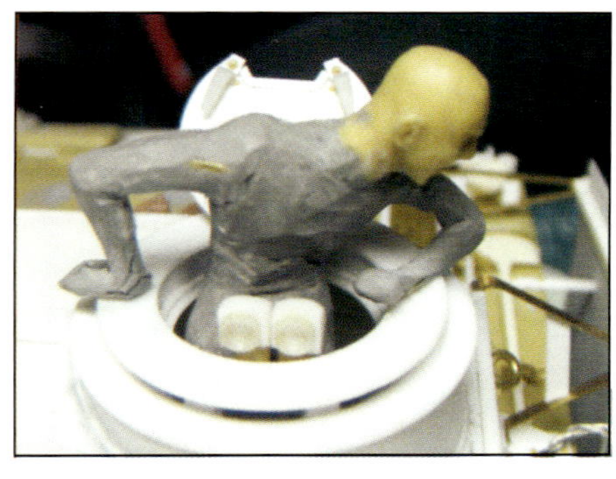

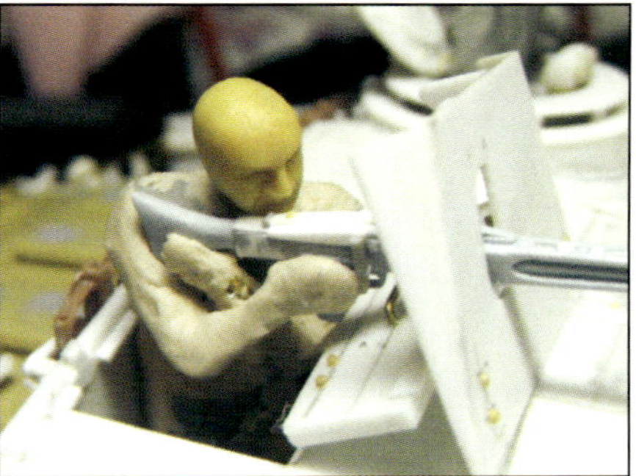

Above: The three figures used some parts of the Nichimo 1/30 scale infantry set but were for the most part built up from putty, plastic card and wire. The MG 42 is an Airfix 1/32 scale piece that has had the overall length extended and extra details added. The images above show the various stages from the initial fitting into the vehicle to the final painting. Below: The fully assembled vehicle, less the tracks and Schürzen, painted in a generic three-colour scheme.

The finished model painted with Tamiya Acrylic Colors and treated with artists oils and various weathering products. The base was made from a polystyrene block and epoxy resin was used to create the water.

Another view of the completed model here on its scenic base and photographed in a realistic setting. All the markings are waterslide transfers taken from Nichimo and Dragon Models kits.

The Sturmgeschütz III assault gun has been a constant favourite with modellers since plastic armour kits became readily available in the 1960s and indeed one of the best-known kits from this early period, Airfix's StuG III Assault Gun in 1/76 scale, is still available today. The shortcomings of many early kits are all too obvious today but they should perhaps be judged only after considering the limitations of the period, both in the technology of mould making and in the paucity of reliable research material. The 1970s was something of a Golden Age for plastic model collectors and builders, with firms in Japan and Europe offering relatively accurate replicas at a reasonable price and it was also at this time that Tamiya, after some experimentation, decided that its armour models would henceforth all be produced in 1/35 scale, an area of plastic modelling that the company controlled for some time. But it was in 1989 that Gunze Sangyo released the first of what would be called multi-media kits, a 1/35 scale Sturmgeschütz III ausf G, which contained injection-moulded plastic parts, photo-etched details, individual track links and other additional options. These pioneering efforts paved the way for the highly-detailed and accurate replicas on offer today. Models of the Sturmgeschütz III are today produced in a number of scales from tiny 6mm wargames miniatures to huge, radio-controlled 1/6 scale versions but these are at the extreme ends of the market and I have chosen to concentrate here, as in previous books, on the most popular modelling scales of 1/35, 1/48 and 1/72. The expansion of the market in recent years has also contributed to the proliferation of smaller, specialised companies producing accessories in resin, photo-etched metal, plastic, wood and other materials allowing a level of super-detailing which had previously been the province of a few master modellers. This list is of course far from exhaustive, mostly due to reasons of space and the almost constant flow of new releases, and should be considered as merely a sampler. An index of manufacturers and their contact details can be found on page 64.

DRAGON MODELS LTD

Dragon Models was one of the first of the major manufacturers to offer accurate replicas in 1/35 scale and their kits are today something of a benchmark. The company's small scale releases are some of the best available. At the time of writing this Hong Kong based company offered twenty-nine different construction kits of the StuG III and the StuH 42, in 1/35 scale and nine kits in 1/72 scale. It should be mentioned that many of these kits utilise common parts and are sometimes re-boxings with different markings and additional components. Dragon Models also offer a number of assembled and pre-painted models of the StuG III in 1/72 scale.

Below: Dragon Models' 1/35 scale StuG III ausf G built with the Eduard Model Accessories photo-etched brass set.

Below: Details of Dragon Models' 1/35 scale StuG III ausf G including the interior and breech of the main gun.

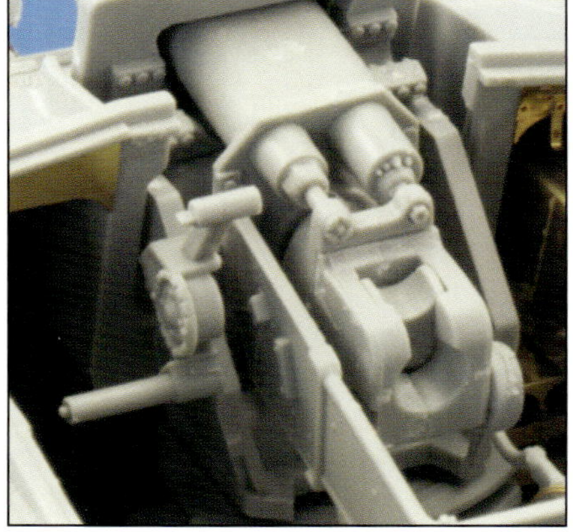

TAMIYA INCORPORATED

This Japanese company, which began life as a sawmill, has been manufacturing plastic kits since 1959, releasing their first armour model two years later. By the early 1970s Tamiya was producing a large range of armoured vehicles and accessories and the company was almost solely responsible for the rise in popularity of 1/35 scale. Like Airfix, the adoption of a constant scale proved popular at a time when many manufacturers were making models to fit the box in which they would eventually be marketed. In addition, Tamiya's early kits were, for their day, highly detailed, relatively easy to assemble and, most importantly, affordable. The 1/35 scale StuG III has been upgraded since its initial release in 1995 and now includes crew figures, optional parts and a detailed interior. Tamiya also offers 1/48 scale versions of the early and late variants of the StuG III.

At left: The 1/48 scale early StuG III ausf G and below that the mid-production version StuG III in 1/35 scale. At right: A detail of the 1/48 scale StuG III hull and superstructure with details by Hauler.

At right: Tamiya's StuG III ausf G in 1/48 scale built and painted according to the kit instructions as a vehicle of Sturmgeschütz-Brigade 237.

TRUMPETER/HOBBY BOSS

Often thought to be separate entities, these Chinese companies are in fact owned by the same corporation, are located at the same address and are essentially the same firm although different models are marketed separately under the two labels. Although the company now seems to be concentrating on modern subjects, Trumpeter does offer a number of StuG III models. Shown below are, 1. The newly-released 1/16 scale StuG III. 2. The StuG III ausf F. 3. Part of the StuG III ausf F hull which is also used for the other early variants produced by Trumpeter.

1

2

3

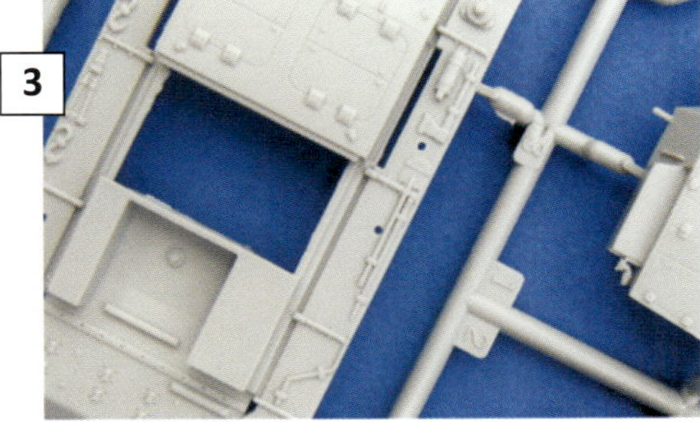

TAKOM

One of the newest model manufacturers, this company started operating in Hong Kong in 2013 as Takom World and shortly afterwards expanded to open a plant in mainland China where all the models are currently produced. First released in 2020, the company's 1/35 scale StuG III ausf G Late Production kit provided the basis from which a range of assault gun models followed including versions of the StuH 42 and early variants such as the StuG III ausf F and F/8 and a kit which contains parts to build either the 7.5cm or 10.5cm howitzer-armed vehicles.

Above, from left to right: The Takom 1/35 scale later-production StuH 42 built from the 2-in-1 kit. The StuG III Early Production model, that is pre-Zimmerit, with Winterketten extended tracks. The StuG III or StuH 42 Early Production 2-in-1 kit in 1/35 scale.

BORDER MODEL

Situated just outside Shanghai, this company began releasing kits based on the StuG III chassis in 2022, the latest offerings being released shortly before this book went to print. The company also produces a range of photo-etched details and workable tracks. The range of assault gun models contain crew figures, photo-etched parts, metal gun barrels and a tool to inscribe the Zimmerit texture. Each kit contains a number of marking options and instructions on camouflage patterns.

Above, left to right: Border Model's 1/35 scale Late Production StuG III. Although this is marketed as a separate kit, I could find no appreciable difference between this offering and the company's StuG III Final Production model which was in fact Border Model's first release. The Late Production StuH 42 and the Early Production model. The latter is essentially the Late Production variant without the Schürzen. All models feature different marking options.

MINIART CREATIVE MODELS

Based in Ukraine, this company has been producing 1/35 vehicles, figures and aircraft models since 2001. The company's current catalogue contains an impressive array of models of the StuG III and StuH 42. All the kits include photo-etched brass parts, extensive marking options, individual track links, with a jig to help with assembly, and extensive interior detailing. The company also offers an extensive range of 1/35 scale figures and diorama accessories.

At left: Miniart's StuH 42 Late Production model with the StuG III Alkett Production October 1943 kit, both in 1/35 scale.

At right: The Alkett 1945 production StuG III model complete with five Pilze sockets, remotely-controlled MG 34 and Series II Schürzen.

REVELL

This company has been producing plastic models for almost as long as Airfix and was at one time known for its highly detailed and accurate replicas. The current catalogue contains a 1/35 scale StuG III ausf G and a StuG III ausf F/8, both of which are early Dragon Models kits released under Revell logo. The company's 1/72 scale StuG III was first released in 2003 but it is still a fairly accurate and reasonably priced representation. It was upgraded with some extra parts, including hard plastic tracks, and new marking options in 2013 and released as a late production vehicle.

Above, left to right: Revell's late production StuG III in 1/72 scale. A more recent re-issue of the late production kit with new markings.

Above: Revell's 1/72 scale late production StuG III built with the upgrade set from Black Dog.

AIRFIX

No listing of model replicas of the StuG III would be complete without at least some mention of the venerable Airfix 1/76 scale kit. First released in 1962, the packaging and box art has changed many times over the intervening years, as has the ownership of the company, although the model itself has remained completely unaltered, including the marking options. The soft plastic tracks, which generations of modellers have complained about endlessly, could only be described as rudimentary and are impervious to any glue. But it was, for its day, an accurate representation of an important armoured vehicle and the starting point for many armour modellers.

Below: Airfix's 1/76 scale StuG III, or 75mm Assault Gun, built recently for the company's catalogue. At left: One of the many re-packagings of this kit, shown here is the 1963 version. By 1965 the plastic bags had been replaced by cardboard boxes.

HAULER

Based in the Czech Republic, this company produces accessory sets in photo-etched brass and resin in 1/48, 1/35 and even 1/87 scale, a size which complements HO scale railway models. A number of items in the company's extensive range are now produced under the Bren Gun logo. Shown below are parts of the photo-etched brass and resin upgrade sets produced to complement the Tamiya 1/48 scale StuG III including hull Schürzen, loader's machine-gun shield, engine grills and rear hull stowage rack. Also visible are the very finely detailed tool clasps and buckles.

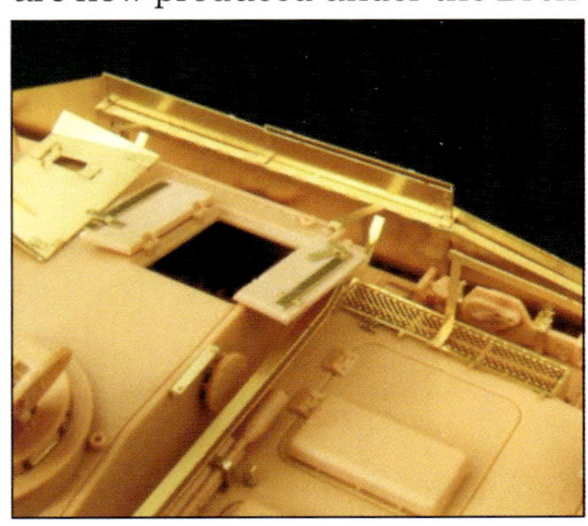

GRIFFON MODEL

Based in China, this company produces a large range of photo-etched brass sets for the StuG III in 1/35 scale to fit most manufacturers' kits, although they do recommend the models of Tamiya and Dragon. Some sets are quite basic, for example containing a hinged machine-gun shield and ammunition drum, while others are extremely comprehensive and are made up of hundreds of brass and resin parts. The photographs below all depict details of the company's Premium Edition upgrade set for an Alkett-produced StuG III in 1/35 scale.

BLACK DOG

Based in the Czech Republic, this company produces a large range of detailed resin accessory and upgrade sets for armoured vehicles in 1/35, 1/48 and 1/72 scale and a small number of complete kits. Shown below are, 1. and 2. The accessory set for Dragon Models' 1/72 scale StuG III ausf G. 3. and 4. Resin details for the Tamiya 1/48 scale StuG III ausf G. 5. The 1/48 scale Tamiya StuG III ausf B built with Black Dog details. 6. The resin set for a 1/35 scale StuG III ausf C/D.

1

2

3

4

5

6

E.T. MODEL

This relatively new manufacturer from China produces upgrade sets in 1/72 and 1/35 scale in brass and resin. Most sets are specifically designed, according to the company, to fit Dragon or Tamiya kits but most would be appropriate for other manufacturers' kits. The company also offers a large number of diorama pieces including fuel drums and canisters, various types of baggage, photo-etched barbed wire and even garden furniture and leaves in 1/35, 1/48 and 1/72 scale. Recent releases from Meng-Models have included photo-etched parts from E.T. Model

Above, left to right: The Dragon Models' 1/35 scale late-production StuG III kit finished with the E.T. Model upgrade set. The mid-production Alkett StuG III, also in 1/35 scale. Dragon Models' 1/72 scale Stug III built with E.T. Models photo-etched brass details including Zimmerit coating.

EDUARD MODEL ACCESSORIES

This Polish company has been manufacturing high-quality accessories in photo-etched brass and resin for many years and was one of the earliest firms to offer upgrade sets in the smaller scales. In addition to the upgrade sets Eduard producess other accessories including metal gun barrels, scale rifles and machine guns, ammunition crates, paint masks and markings.

Above, left to right : Details of the 1/35 scale upgrade set showing the shield for the remote-control machine gun, internal hatch parts and cupola periscopes. The so-called Waffle pattern Zimmerit indicitive of Alkett-built vehicles applied to Tamiya's 1/48 scale StuG III.

VOYAGER MODEL

Voyager have been manufacturing comprehensive upgrade sets for scale models since 2003 with the release of their first set for 1/35 scale armour. The company also produces turned aluminium barrels with muzzle brakes for both the late and early 7.5cm main guns of the StuG in 1/35 scale and accessories in resin including a detailed commander's cupola. Shown below are, from left to right, the late Schürzen in 1/35 scale, parts of the detail set for the Dragon Models 1/35 scale StuG III and photo-etched brass and resin details applied to the Tamiya 1/48 scale kit.

ABER

This Polish company has been manufacturing and selling upgrade sets since 1995, working in photo-etched brass, milled aluminium and brass, stainless steel and even wood. For some time now Tamiya has included Aber products with a number of their models. A full list of the accessories made specifically for the StuG III and StuH 42 in all scales would be far too large to reproduce in this book and the reader can find the company's contact details on page 64. Shown here are, 1. A late pattern muzzle brake for the 7.5cm L/48 gun in 1/35 scale. 2. An early type muzzle brake and barrel in 1/35 scale for the StuG III. 3. A selection of 7.5cm projectiles, including armour-piercing and high explosive rounds, also in 1/35 scale. 4. A 1/72 scale 7.5cm muzzle brake and barrel for a StuG III. 5. A standard German fire-extinguisher and bracket in 1/16 scale. 6. An antenna insulator and antenna in 1/16 scale. These last two items would be appropriate for most German armoured vehicles.

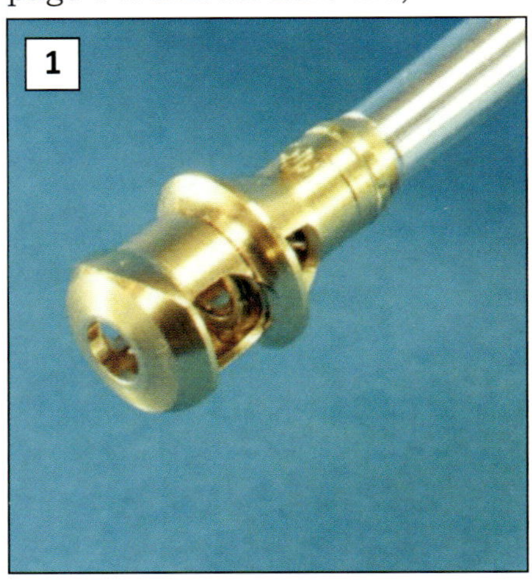
1

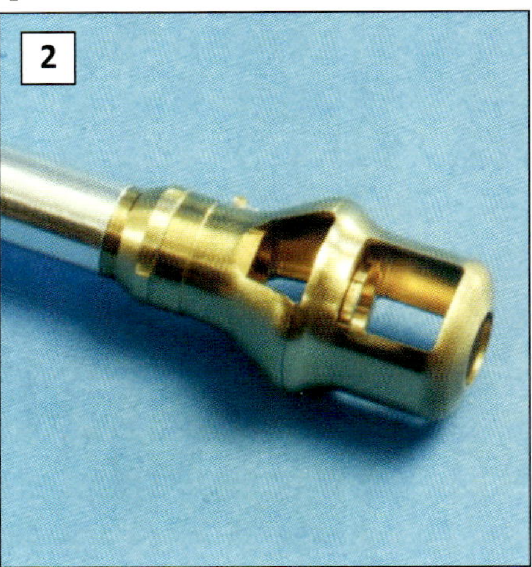
2

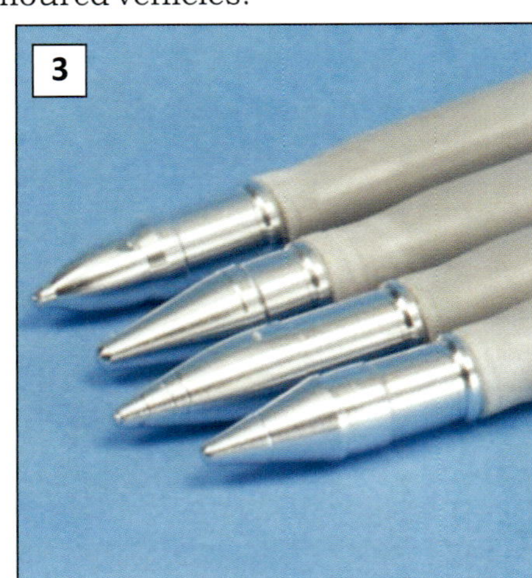
3

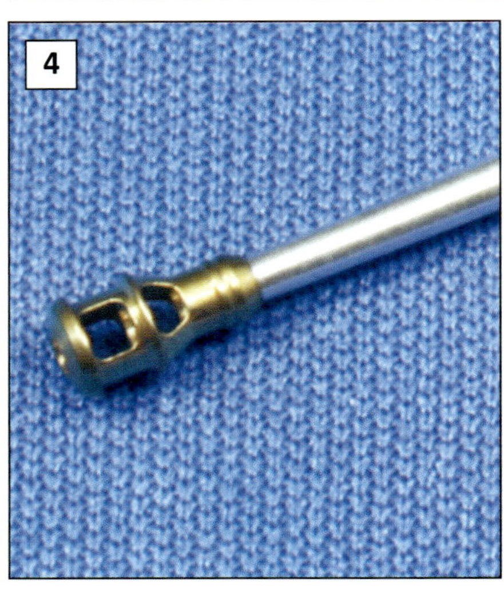
4

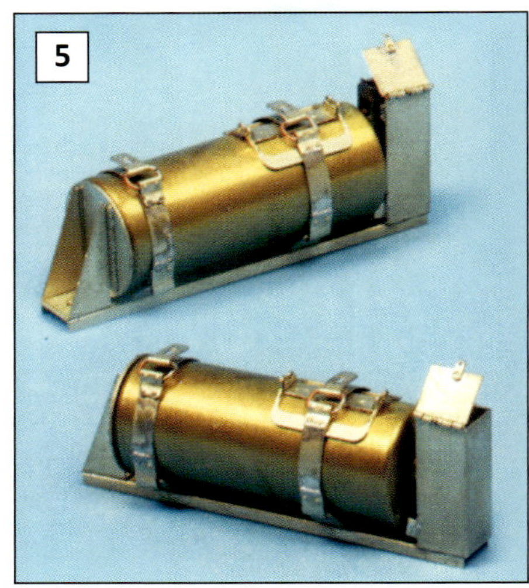
5

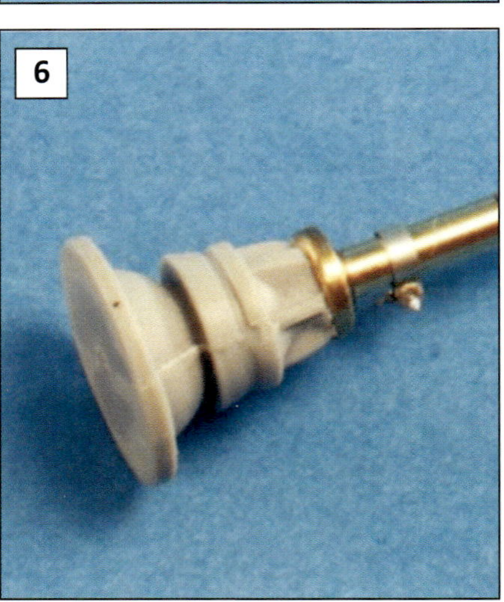
6

AFTERMARKET TRACKS

Most plastic kits today contain realistic tracks made from a form of vinyl which, unlike the older types, can be glued together, or hard plastic individual links. Nevertheless, there are many modellers who wish for the highest level of detail and realism possible and consequently the number of quality aftermarket products available seems to have increased in recent years, despite the extra expense and work required. Motorised models require workable tracks made up from individual links held together by pins, as were the full-size items. The examples shown here, all in 1/35 scale, are 1. StuG III tracks in white metal from Russian company Master Club. 2. Early StuG III tracks in plastic from Kaizen. 3. Individual early StuG III track links in plastic from Miniart.

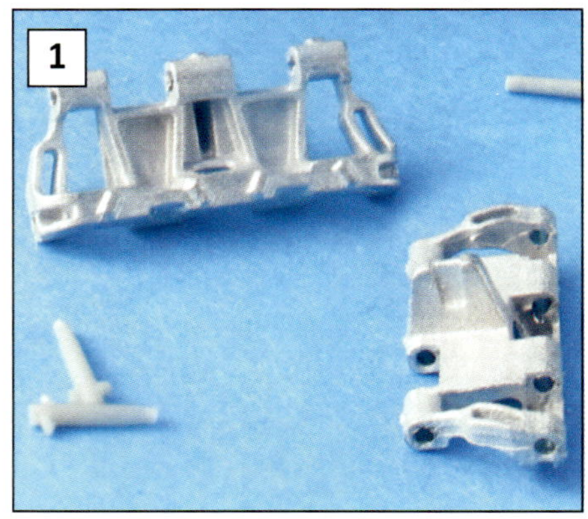
1

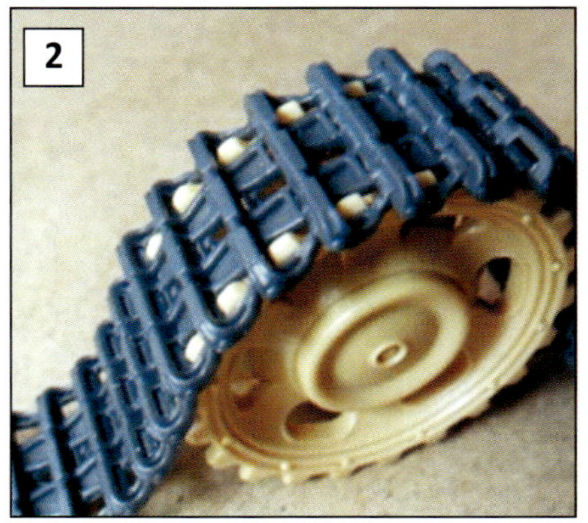
2

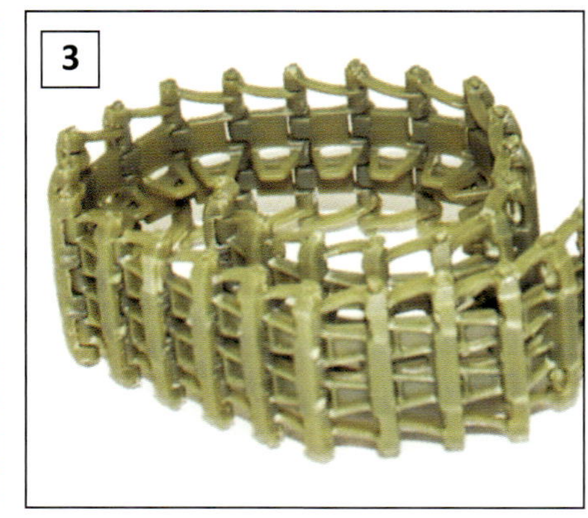
3

The major production version encountered during the period covered by this book was the Sturmgeschütz III ausf G model of which over 8,000 examples were assembled between December 1942 and April 1945. This variant far outnumbered all the earlier models combined and I have in the main restricted the modifications listed below to that vehicle. Many of the earlier models, particularly the Sturmgeschütz III ausf F and Sturmgeschütz III ausf F/8, did soldier on until the end of the war and various technical aspects and identifying features of those models are shown and discussed in the Camouflage & Markings section which begins on page 17.

The principal manufacturers of the Sturmgeschütz III ausf G were the firms of Altmärkische Kettenwerke (Alkett) of Berlin-Borsigwarde and Mühlenbau und Industrie Aktiengesellschaft (MIAG) of Brunswick. Maschinenfabrik Augsburg-Nürnberg (MAN) supplied the last 142 of their Pzkpfw III chassis which were converted to assault guns. The Sturmgeschütz III was armed with the 7.5cm StuK 40 L/48 gun and these were manufactured by Wittenauer Maschinenfabrik of Berlin-Borsigwald and the Skoda Works at Pilzen and complete weapons were delivered to the assembly plants. The muzzle brakes mentioned below could, and were, fitted to either manufacturer's weapons. The 10.5cm StuH 42 L/28 guns for the Sturmhaubitze, a modified version of the leichte Feldhaubitze (leFH)18, were produced by Menck & Hambrock Maschinenfabrik of Hamburg. The initial muzzle brake was sometimes replaced by the so-called finned version of the leFH 18/40 but I have been unable to establish with any certainty if this was due to factory shortages, a field modification or a production upgrade.

Although I have endeavoured to make this list as comprehensive as possible, limitations of space have dictated that some small detail changes have been omitted and the reader should note that I have not included internal technical alterations. It should also be noted that when earlier production models of the Sturmgeschütz and Sturmhaubitze were returned to Germany for major repairs many of the latest features and modifications were added and similarly it is not uncommon to see vehicles assembled at Alkett with parts which would normally identify a MIAG-built assault gun and vice versa. Throughout this section I have used the abbreviations HWA for Heereswaffenamt, the army's ordnance office, Fgst.Nr. for Fahrgestelle or chassis number, which was unique to each vehicle, StuH for Sturmhaubitze and StuG for Sturmgeschütz.

December 1942. Production of the StuG III ausf G began at Alkett with Fgst.Nr. 91651.

The superstructure side plates were set at a steeper angle than previous models and a pistol port was added to the left hull near the driver's position in place of the vision slit. The gunner's Sfl.ZF1a periscope sight protruded through an aperture in the roof of the crew compartment as it had in the earlier versions.

An armoured shield, hinged at the bottom, was mounted in front of the loader's hatch to accommodate an MG 34 machine gun. These shields were also retro-fitted to StuG III ausf F/8 models from early 1943.

On the first examples the fume extractor fan was mounted on the superstructure roof next to the commander's cupola. The latter was fitted with seven periscopes which faced towards the rear and to the sides with a blind spot at the front which was covered by a metal plate. Forward vision was provided by the SF.14Z Scherenfernrohr binocular rangefinder with which each vehicle was equipped.

The inner section of the cupola, comprising the hatch and hinge arrangement, could be rotated. A rubber stopper mounted on a metal bracket was fitted to the hatch which held it at 45 degrees when opened. At the forward edge of the hatch a small hinged flap allowed the binocular rangefinder to be used when the hatch was closed.

The towing cable clasps were moved further forward on the trackguards and the long-handled shovel and towing hooks were relocated.

The single-piece transmission access hatches were replaced by two-piece versions which opened outwards. Generally speaking, the hatches which incorporated the locks were placed on the outer side hatch on Alkett-built vehicles and on the inner side for MIAG, although the actual hatches were interchangeable.

January 1943. The fume extractor fan was relocated from the roof to the rear outside wall of the fighting compartment. A sliding armour guard was added to the roof to protect the Sfl.ZF1a periscope gunsight.

Production of the StuG III ausf G began at MIAG with Fgst.Nr. 95001 but the first ten vehicles were not fully assembled until the following month.

A HWA report referring to Pzkpfw III production for January 1943 mentions a total of forty-six vehicles as earmarked for conversion to StuG III. Of these, thirty-four had been assembled by MNH, eleven by MAN and one by Henschel. That they are described as being armed with the 5cm L/60 gun would indicate that they were fully built tanks and not just a hull with suspension.

February 1943. Smoke grenade dischargers, or Nebelwurfgerät, were fitted on each side of the superstructure at the front. Vehicles assembled at MIAG at this time had the fire extinguisher mounted at a 45-degree angle.

A new base coat, initially referred to as Dunkelgelb nach Muster (1), was introduced in this month. There was no change in the primer coats which were dark grey for the crew compartment and gun and red for the chassis.

..........text continued on page 51

Notes

1. Later reclassified as Dunkelgelb RAL 7028.

An Alkett-built StuG III ausf G assembled in January or February 1943. Technical features mentioned in the text include: A. The early commander's cupola with blind spot at the front. B. The Sfl.Zf1a periscopic gunsight. C. Cut-out in the armour plate over the driver's visor. Note the two holes for the driver's KFF2 periscope which was dropped from production in March 1943. D. 30mm Zusatzpanzer bolted-on armour. E. Extended hull towing eyes.

Photographed in southern France in late 1943, this StuG III ausf G is one of the vehicles built on later-production Pzkpfw III chassis as evidenced by the single-piece transmission access hatches (F). Note also the shield (G) for the MG 34 with its distinctive pear-shaped aperture (H) for the gun and the single-piece bolted-on armour over the driver's visor (I). Just visible is the support rail for the Series I Schürzen (J) and the trackguard support (K) featuring a triangular brace indicative of MIAG-assembled vehicles.

.........text continued from page 49

March 1943. The KFF2 driver's periscope was dropped from production which meant that the gap in the bolted-on additional armour plates above the driver's visor was no longer required. As a number of these plates had already been fabricated a metal insert was used to fill the gap and welded in place. Units in the field were also ordered to effect this change. By April a single-piece version was standard.

The first production series examples of the StuH 42, based on the ausf G chassis and superstructure, were assembled by Alkett (1) which was the sole manufacturer of this vehicle. A separate range of Fgst.Nr. was not allocated and numbers from the StuG III range were used.

At MIAG, a programme of rebuilding MAN-built Pzkpfw III ausf M tanks as StuG III ausf G assault guns began which would see a total of 127 conversions completed by the following November. These vehicles retained the deep wading muffler and single-piece transmission inspection hatches.

April 1943. In place of the composite 50mm and 30mm Zusatzpanzer armour plates bolted on to the hull front, a single 80mm-thick plate was introduced into the manufacturing process. Due to a shortage of steel plate, this change was not fully implemented until the following November and some vehicles were fitted with the bolted-on armour until that time. Armoured skirts or Schürzen, referred to throughout this section as Series I (2), were fitted to both sides of the hull as a defence against Russian anti-tank rifles. By the following month, units in the field had been ordered to attach Schürzen to their assault guns. StuG III ausf F/8 models were also retro-fitted with armoured skirts.

At some time after the introduction of the Series I Schürzen, it is difficult to be precise, supports were added to the track guards to accommodate the extra weight. The supports fitted by Alkett were of pressed metal and semi-cylindrical in shape while those used by MIAG comprised a metal tube with a triangular brace, sometimes referred to as a fin. The Series I Schürzen are examined in more detail on page 56.

Also in April, Pzkpfw III ausf J and Pzkpfw III ausf L chassis held at MIAG were converted to StuG III assault guns. By December 1943 a total of 100 conversions had been completed.

May 1943. The Nebelwurfgerät smoke grenade dischargers were dropped from production. Photographic evidence would suggest, although not confirm, that this change was gradual and they are seen on vehicles known to have been assembled in June.

A new muzzle brake, fitted with side deflector flanges, was introduced into production. It should be remembered that many modifications were made only after existing stocks of an item had been exhausted and the earlier muzzle brake could be seen for some time.

Notes

1. A trial series of twelve vehicles based on rebuilds of earlier models had in fact seen service in November 1942 near Leningrad.

2. The terms Series I and Series II were not used during the war and are in fact my own invention, employed here purely as a matter of convenience.

An Alkett-built StuG III ausf G and Bergepanzer recovery vehicle, both fitted with Ostketten tracks, photographed on the Eastern Front. This image was probably made during the late summer of 1944 and points to note include the Series II Schürzen, Topfblende mantlet and the travel lock for the 7.5cm gun. Vehicles were fitted with the mount for the Rundumsfeuer remotely-controlled MG 34 from May 1944 but the guns and their shields were not available until the following August. Although it is difficult to see here, the Zimmerit has been applied in the so-called waffle pattern.

A StuH 42 photogphed on the Eastern Front, probably during the late spring of 1944. Note the enlarged aperture of the machine-gun shield which was introduced in January or February of the same year to accommodate the MG 42. Note also the field-modified Schürzen which pivot on a metal bar welded to the trackguard. The small pennant in the foreground identifies Grenadier-Regiment 434 which fought in the battles around Kovel at this time and Sturmgeschutz-Brigade 190, which also took part in the fighting in the same area, is usually credited with the invention of these swinging side skirts.

September 1943. Simplified drive sprockets without the covers for the retaining bolts were incorporated into production at MIAG. This modification had in fact been made by units in the field for some time. Alkett continued fitting the cover until December.

During the final stages of assembly, a thick paste, usually referred to by its commercial name of Zimmerit, was applied to all vertical armour surfaces. The paste was applied with special tools which gave a distinctive pattern of ridges and hollows.

Generally speaking, vehicles assembled by Alkett left the plant coated Zimmerit in the so-called waffle pattern and those produced by MIAG had a pattern of small squares (1). It was not until January 1944 that units in the field were ordered to apply Zimmerit to their assault guns with, as photographic evidence shows, very mixed results. This is hardly surprising as the application was quite a complicated process.

It may have been at this time, or slightly earlier, that a modified trackguard support of simpler construction was incorporated into production at Alkett.

October 1943. The commander's cupola was reinforced by an armoured deflector which was welded in place. Many vehicles left the MIAG assembly line without this modification until well into 1944.

November 1943. A cast gun mantlet, referred to as a Topfblende (2), was introduced into the assembly process at Alkett. The earlier box-type mantlet continued to be installed at MIAG until the end of production but a number of that firm's assault guns were later rebuilt at Alkett and these could have been fitted with the Topfblende.

Cast-steel return rollers replaced the earlier, rubber-tyre versions. Stocks of the new return rollers were often unavailable and the rubber versions can be seen well into 1944.

The hull rear plate was now interlocked whereas the plates on previous vehicles had been butt welded.

Beginning in November a shortage of ball bearings meant that the commander's hatch on most, but not all, vehicles was welded in place where it had previously been capable of rotating (see December 1942). In addition, the seven periscopes were now arranged with one facing forward while the blind spot was now at the rear and no longer covered by a protective metal plate. The rubber stopper mounted on the hatch was retained.

December 1943. Drive sprockets without the covers for the retaining bolts were incorporated into production at Alkett.

A canvas rain and dust cover could be mounted behind the mantlet of the main gun but it is rarely seen in contemporary photographs.

Notes

1. Photographs, taken on both the Eastern and Western Fronts, show patterns of horizontal ridges which are are so similar that I am reluctant to believe that the paste has been 'field-applied'.
2. Although it should not need repeating, the term Saukopf or sow's head, was not used to describe these mantlets during the war.

Early 1944. Early in the year, or perhaps late in 1943, additional armour in the shape of concrete reinforcement was added to the superstructure by individual units. That this was a field modification and not applied in the factories is fairly certain and indeed photographs exist of troops applying the concrete. Most often, steel bars or similar were first welded to the area onto which the concrete was poured and then smoothed. Endless variations existed but most commonly the concrete extended over the front plate of the superstructure and around the driver's visor. It could also be added to the exposed side of the commander's cupola and the sides of the superstructure, although this practice was less common.

January 1944. A bracket was fitted to the rear wall of the crew compartment to secure the engine inspection hatches in the open position.

A simplified towing pintle was introduced into production.

In January, or possibly February, the aperture of the MG shield was changed from a pear shape to a large rectangle to accommodate the MG 42.

March 1944. An improved mounting system for the hull Schürzen, referred to here as Series II, was introduced which allowed the plates to be fitted in combination with Ostketten extended tracks. This modification is examined in detail on page 58.

On some, but not all, vehicles the starter handle was moved to the hull rear.

Between March and May 1944 a total of fifty-seven Pzkpfw III ausf J and Pzkpfw III ausf L were converted to StuG III ausf G at MIAG. These vehicles can be identified by the escape hatch which was retained on the hull side. Although undocumented, photographic evidence suggests that a number of these Pzkpfw III tanks were converted at Alkett at about this time, the date confirmed by the presence of the Series II Schürzen.

April 1944. The bolted-on armour plate in front of the loader's position was replaced by an 80mm-thick plate which was welded in place.

May 1944. A remotely-controlled MG 34 machine gun, referred to as a Rundumsfeuer, replaced the hinged armour shield in front of the loader's hatch. The new gun, which incorporated a small shield, was held in a purpose-built cradle and fixed to a cylindrical mount and the loader's hatch was re-designed to open towards the sides in two halves. But due to a shortage of guns, vehicles were produced for several months with just the mount which was covered by a metal plate held in place by six bolts. It is possible that none of these guns were fitted at MIAG until September 1944 or later.

A close defence weapon, referred to as a Nahverteidigungswaffe, was to have been mounted in the roof but again, due to a shortage of weapons, the installation was delayed and the resulting hole in the roof was covered by a bolted-on metal disc.

A new muzzle brake featuring a round rear baffle was introduced.

An example of concrete armour as mentioned in the text. Note that metal reinforcements have been welded in place just above the driver's visor. Additional armour plates have also been fitted in front of the commander's cupola. The Schürzen are of the Series I type.

Notes

1. The modifications to the cupola mentioned here were actually ordered into production in January 1944 but it is fairly certain that they were not implemented until May or even later.

2. These Schwingschürzen, and their official acceptance, are mentioned in an award document which names the battalion's Oberwerkmeister Kurt Hanel as the inventor.

At both MIAG and Alkett a segment was cut from the rear section of the upper ring of the commander's cupola to allow the hatch to lay horizontally when opened and the large rubber stopper was replaced by a flat wooden version. Photographic evidence shows that this modification had been employed by a number of front-line units for some time but troops in the field were not ordered to make the alteration until November 1944. Also in May, or perhaps slightly later, a wire loop handle was fitted to the inside of the hatch which allowed the commander to close the hatch quickly in an emergency (1).

June 1944. A travel lock for the main gun was added to the hull front. This modification was ordered to be carried out by units in the field in the following month.

A tube for a coaxial machine gun was added to the box-type main gun mantlet and three distinct variants are seen. In the first, the tube protruded slightly and the arrangement of the mantlet's bolts was unchanged. With the second version the tube is not visible, only the hole through which the gun was fired, and the right hand bolt on the upper edge was moved higher. The third type is seen with field-modified gun mantlets where the tube is not visible but the position of the bolts was not altered.

At about this time a standardised stowage rack for the hull rear deck was introduced and the S-shaped towing hooks were replaced by C-shaped versions.

Supplies of the Nahverteidigungswaffe were finally available.

August 1944. An order of 19 August directed that tanks and assault guns were to be painted in a standardised colour scheme before leaving the factory. The base colour of Dunkelgelb RAL 7028 was retained but over this, large patches of Olivgrün RAL 6003 and Rotbraun RAL 8017 were to be painted. In addition, interiors would no longer be painted in Elfenbein RAL 1001 but left in the primer colour.

Vehicles began leaving the assembly lines fitted with the Rundumsfeuer remotely-controlled MG 34 but shortages dictated that this feature was not universal until the beginning of October.

September 1944. Three sockets, or Pilze, were welded to the roof of the fighting compartment allowing for the temporary mounting of a 2-ton jib boom that could be used to remove heavy parts such as the engine.

The assembly firms were ordered to cease the application of Zimmerit paste with immediate effect.

It is possible that a limited number of a redesigned commander's cupola was introduced at this time (see December 1944).

The muzzle brake of the 10.5cm gun of the StuH 42 was no longer fitted and from this time the gun included a reinforced recoil system. This meant that it could no longer be used with the Topfblende mantlet nor was there space for the co-axial machine gun and the early welded mantlet was reintroduced.

The HWA approved the development of a new type of Schürzen which was based on the so-called 'swinging side skirts' pioneered by Sturmgeschutz-Brigade 190 examples of which are shown in the Camouflage & Markings section on page 22 (2).

A late production StuG III ausf G, probably photographed in the West, showing the standard stowage rack which was introduced into production in mid-1944. This Alkett-built vehicle is also fitted with the Rundumsfeuer remotely-controlled machine gun and has the five Pilze socket arrangement.

Photographed on the Eastern Front late in the war, all three assault guns visible here are fitted with field-modified spaced armour which in the case of the nearest vehicle is partly coated in Zimmerit. Note the raised hatch of the commander's cupola with the wire loop handle. Also of note are the shield of the Rundumsfeuer machine gun, Topfblende mantlet and gun travel lock.

October 1944. A coaxial machine gun was added to the Topfblende gun mantlet. As existing mantlets could not be re-worked to accept the co-axial machine gun a complete redesign was necessary and the new Topfblende was assymetrical, featuring a distinct bulge on the left, that is the driver's side, to accommodate the machine-gun tube. It is highly likely that this mantlet was fitted to Alkett-built vehicles only (1).

A new muzzle brake was introduced with two round baffles but this change was never fully implemented and many vehicles left the factories equipped with the old versions

On some vehicles a rack was welded to either side of the crew compartment to hold spare track links.

The additional armour of the two centre plates of the Series II Schürzen was no longer fitted. It was reported that Ostketten tracks could no longer be employed by these vehicles and this presumably meant that the U-shaped mounting brackets were altered (see also page 58).

In late October or early November, the HWA ordered that the use of Dunkelgelb RAL 7028 as a base colour was to cease. Armoured vehicles would be camouflaged at the assembly plants in patches of Olivgrün, Rotbraun and Dunkelgelb applied over the primer colour (2). This scheme may have been short lived.

At Alkett, vehicles assembled in this month were fitted with five Pilze sockets in place of the previous three. Some sources suggest this modification may have actually been introduced in November.

November 1944. A redesigned, larger towing coupling was introduced.

The front towing eyes of the extended hull sides now featured a small 'tooth' along the bottom edge.

At MIAG, towards the end of the month, the five Pilze socket arrangement was introduced. It would appear that this modification was standard by the following month but it seems that vehicles fitted with three Pilze sockets were not to be modified.

An order dated Tuesday, 21 November 1944 directed that the MG 34 of the Rundumsfeuer was to be replaced by the MG 42. If this order was carried out it would have necessitated a redesign of the complex cradle which held the gun.

On 29 November a new order stipulated that armoured vehicles were to be painted in a base coat of Dunkelgrun RAL 6003 over which patches of Rotbraun 8017 and Dunkelgelb RAL 7028 were to be painted in sharp contours. This scheme was not to come into effect until 1 March 1945.

December 1944. A memo dated 14 December 1944 stated that it had not been possible to equip all 'Sturmgeschütz 40 (Fahrgestell III und IV)' with the rotating hatch which was supposed to have been reintroduced in the previous September (3). The new cupola featured the periscope arrangement of the initial version installed from December 1942 but incorporated the horizontal hatch with the wire loop handle of the May 1944 version.

The exhaust deflector shields were dropped from production, and commonly removed by units in the field, to improve access to the tow coupling

On Wednesday, 20 December 1944 the HWA directed that the factory-applied camouflage scheme outlined in the November instructions was now to come into effect immediately.

Notes

1. Meaning that they were not fitted to rebuilds.
2. A number of accounts have incorrectly interpreted this directive to mean that the primer coat formed part of the camouflage pattern.
3. Disseminated by Chef der Heeresrüstung und Befehlshaber des Ersatzheeres (Chef H Rüst u BdE) the Chief of Army Armaments and Commander of the Replacement Army.

A feature of German tanks and assault guns from mid-1943 was a system of spaced armour plates referred to collectively as Schürzen, literally skirts, and although these were introduced prior to the period covered by this book they were fitted to fully-tracked armoured vehicles in some form until the end of the war. Early experiences on the Eastern Front had shown that the hull and turret sides of the Pzkpfw III and Pzkpfw IV tanks and the Sturmgeschütz III assault guns, could be penetrated by the Red Army's 14.5mm PTRD and PTRS anti-tank rifles. Soviet trials had shown that these light, man-portable weapons were effective at ranges of up to 800 metres and even the Panther, when it entered service, was susceptible to these rifles which were usually deployed en masse. The commander's cupola of the Sturmgeschütz III was found to be particularly vulnerable. Experiments at the front involving additional armour plates, concrete and spare track links proved less than satisfactory and as early as January 1943 Hitler was presented with the idea of fitting armoured vehicles with spaced armour on which the enemy projectile would expend its energy. Trials were held in February and March using 5mm-thick armour plates and Hitler was so impressed that he ordered that Schürzen be introduced immediately into the production process of the Pzkpfw III, Pzkpfw IV and Panther tank and the Sturmgeschütz III assault gun, anticipating that these vehicles would be ready for the planned offensive on the Kursk Salient. Interestingly, combat trials were undertaken by the companies of Panzer-Regiment 21 in May 1943 and the subsequent report suggested, contrary to earlier findings, that any advantages offered by spaced armour plates were outweighed by the system's shortcomings, of which there were many. But events had overtaken the regiment's negative recommendations and the first assault guns fitted with a full set of Schürzen armour plates had left the assembly lines in April. Indeed, as the front-line trials were taking place, the Generalinspekteur der Panzertruppen reported that more that 1,100 Schürzen assemblies were earmarked for units operating on the Eastern Front with 350 built for the Sturmgeschütz III.

At right, below: A. Sturmgeschütz III ausf G fitted with Schürzen of the initial design which I have referred to throughout this book as Series I purely as a matter of convenience. This system was introduced into the production process in April 1943. The supporting rail (1) for the hull plates was held in place by three V-shaped brackets (2) which were bolted onto metal tabs (3) welded to the roof of the crew compartment and the rear hull. The armour plates were suspended from five hook-like hangers bolted to the trackguard (4) and four smaller hangers (5) fitted to the rail.

B. The same Sturmgeschütz III ausf G vehicle with the Schürzen plates in place. The hangers of the supporting rail (5) and those fitted to the trackguards (4) protrude between each plate.

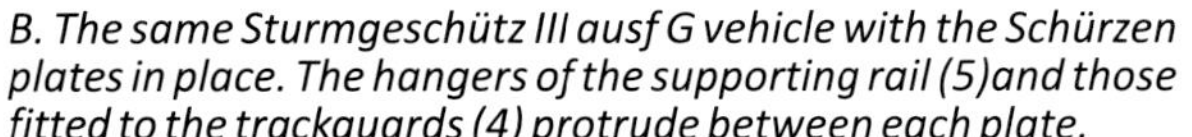

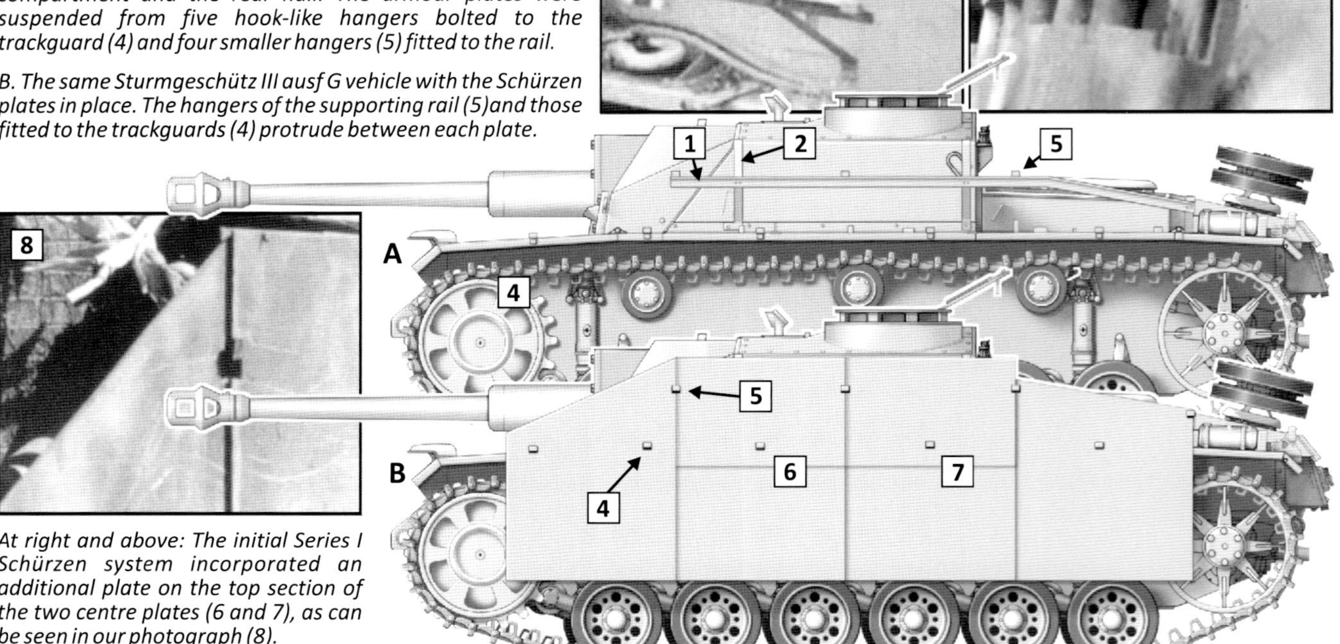

At right and above: The initial Series I Schürzen system incorporated an additional plate on the top section of the two centre plates (6 and 7), as can be seen in our photograph (8).

Sturmgeschütz III ausf G Fgst.Nr.95279 assembled by MIAG in June 1943. The Series I Schürzen system described above is clearly visible here including the method of mounting the additional armour of the two centre plates. These appear to be missing on many Alkett-built vehicles of this period and they may have been simply omitted or fixed to the inner face as they were in the Series II model.

A Sturmgeschütz III ausf G als Befelswagen, the command version of the assault gun, identified by the Sternantenna D for the FuG 8 radio. This vehicle is fitted with the Series I Schürzen of the earliest pattern with the square lower edges of the first and last plates. Note that a number of the hangers protrude well past the surface of the armour plates and have been drilled through to hold the string or wire which is visible here. Although difficult to discern there are traces of Zimmerit applied in the fashion typical of MIAG-built vehicles.

Sturmgeschütz III assault guns of Sturmgeschütz-Brigade 276 photographed in June or July 1943 just a few months after the introduction of the Series I Schürzen. The positioning of the plates tended to push mud into the drive train which placed excessive strain on the engine. This problem was alleviated somewhat by cutting away the lower front corner of the first armour plate as can be seen here on the vehicle closest to the camera. Although a number of sources suggest that this modification was also made to the last plate I have been unable find a single photograph that would confirm this.

Many of the difficulties documented during the May 1943 combat trials conducted by Panzer-Regiment 21 very quickly surfaced once the vehicles fitted with the Series I Schürzen became operational. The supporting rails proved to be too weak to absorb even mild pressure and the armour plates were easily displaced or lost entirely in the muddy conditions of the Eastern Front. Even more seriously, the position of the plates tended to push mud into the drive train which placed excessive strain on the engine. Despite the complications experienced with the initial attempt, the basic concept of spaced armour was deemed to be valid and from March 1944 a new design was incorporated into production which I have referred to here as Series II. This system consisted of a strengthened rail (1) and bracing brackets (2). Triangular hangers (3) replaced the smaller square variants of the Series I Schürzen and the hangers fitted to the trackguards (4) were also redesigned. The forwardmost plate was secured by a triangular-shaped metal tab (5) in addition to the rectangular hanger (6). The plates were of the same basic dimensions as the Series I versions with the lower edge of the first plate cut away but the height of the last plate was increased. U-shaped brackets (7) were fitted to the inner face of the Schürzen plates and these corresponded to the hangers of the support rails and the trackguards. The construction of these brackets allowed the plates to be held at varying distances from the trackguards thus facilitating the use of the wider Ostketten tracks. This had not been possible with the Series I Schürzen. The two centre plates were reinforced on the top half, here shown in a darker shade (8), but the extra plates were now positioned on the inner face and held in place by two bolts (9). Photographs show that the U-shaped brackets were sometimes bolted in place and sometimes welded.

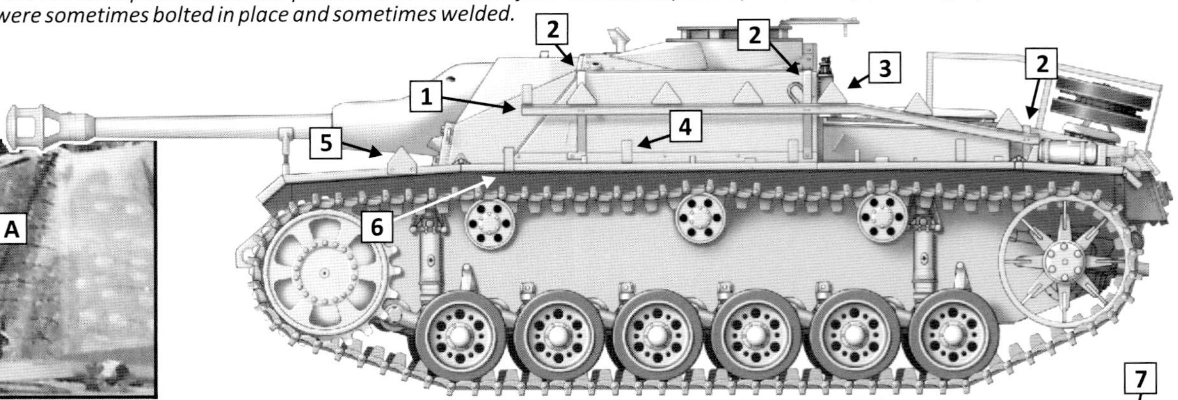

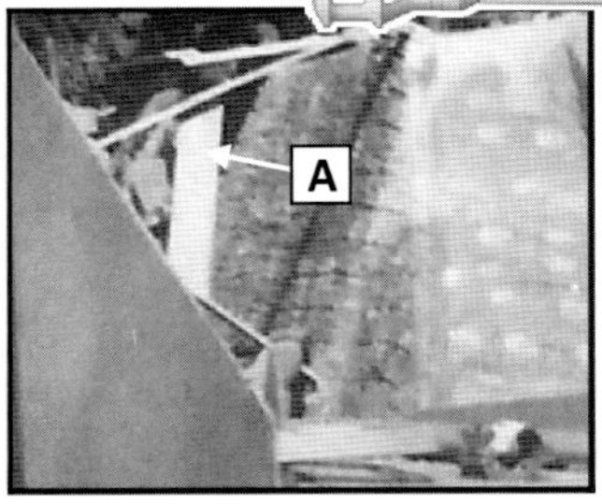

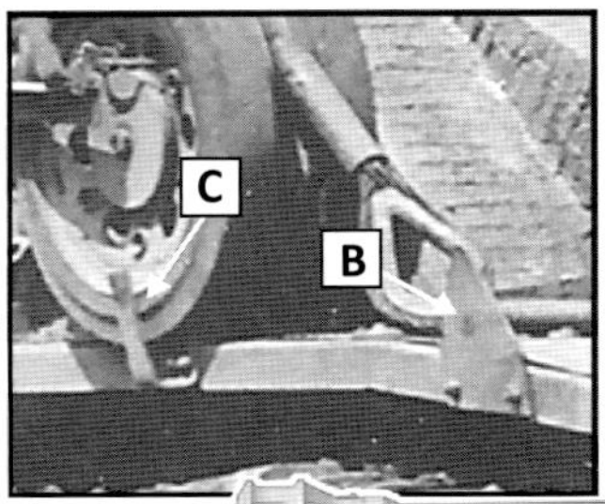

At left: The front bracing bracket of the supporting rail (A). Triangular-shaped tab (B) and hanger (C) bolted to the side of the trackguard. The inner face of the Series II Schürzen showing the triangular hangers (D) and U-shaped brackets (E).

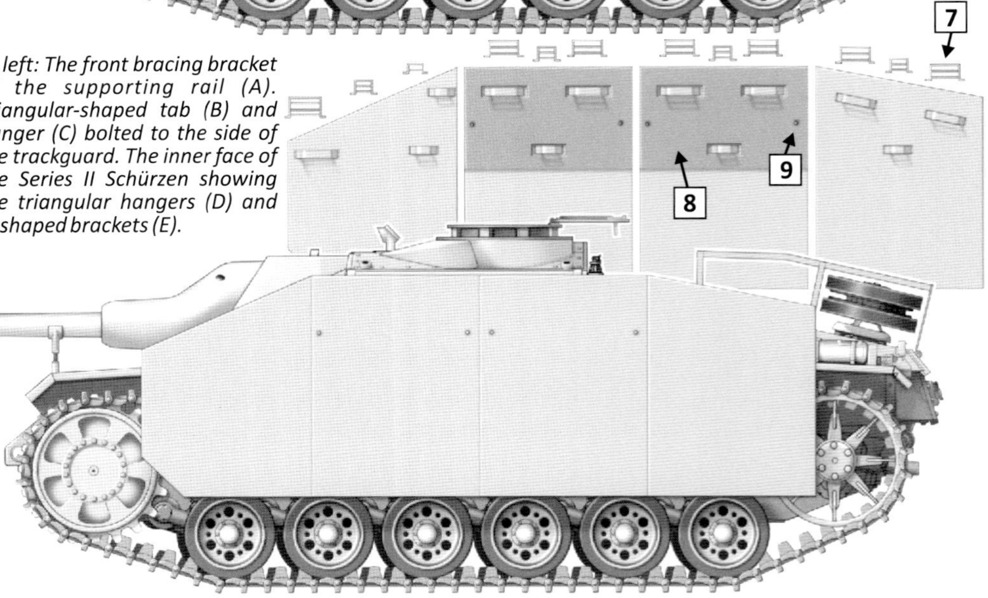

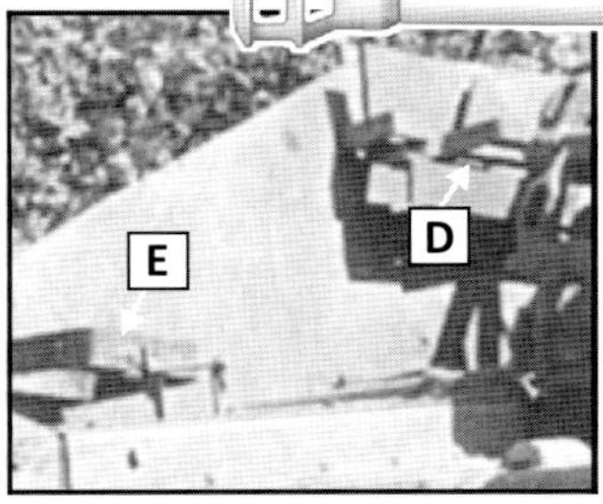

A Sturmhaubitze 42 of Heeres-Sturmgeschütz-Brigade 303 photographed during the summer of 1944. This vehicle is fitted with the Series II Schürzen plates with the U-shaped brackets welded to the inner faces while the image on the facing page shows the version where the brackets were bolted on. On both variants the bolts which held the additional armour of the two centre plates are visible on the outer surface.

An Alkett-built Sturmgeschütz III ausf G photographed on the Western Front in late 1944. The bracing bracket of the Schürzen support rail, a number of the triangular hangers and at least one of the U-shaped brackets are visible here. Note that the latter have very obviously been bolted in place.

A Sturmhaubitze 42 photographed in the East during the summer of 1944. This image provides a rare view of the inner face of one of the centre plates with its additional armour and U-shaped brackets, here welded in place. In the background, one of the trackguard hangers Is being repaired.

STUG III AND STUH 42 PRODUCTION, DECEMBER 1942-DECEMBER 1944

As mentioned earlier the StuG III ausf G was manufactured by the firms of Alkett and MIAG with MAN supplying 142 Pzkpfw III chassis which were completed as assault guns. Alkett was solely responsible for the assembly of the StuH 42. Alkett was initially allocated the Fahrgestell number range Fgst.Nr.91651 to 93850, which probably covered the production run to January 1944, followed by Fgst.Nr.93851 to 94250, Fgst.Nr.105001 to 108800 and Fgst.Nr.108801 to 108920. There was no differentiation between StuG III and StuH 42 with the numbers applied to either type. MIAG was allocated Fgst.Nr.95001 to 97600. It should be remembered that the range of Fgst.Nr. is not necessarily an accurate indication of the number of vehicles assembled and it was usual for the HWA to allocated more numbers than would have been strictly necessary to fulfil a contract as a security measure. The 142 MAN Pzkpfw III chassis were numbered Fgst.Nr.76127 to 76210 and Fgst.Nr.77351 to 77408.

Year	Month	Alkett		MIAG	Total		Remarks
		StuG	StuH	StuG	StuG	StuH	
1942	December	120			120		A total of 129 StuG III accepted by the HWA.
1943	January	130			130		
	February	130		10	140		Production at MIAG had actually begun in January.
	March	144	10	53	197	10	Conversion of 127 MAN-built Pzkpfw III chassis begins at MIAG.
	April	151	34	77	228	34	Conversion of the 100 MIAG-built Pzkpfw III chassis begins.
	May	140	45	120	260	45	
	June	155	30	120	275	30	
	July	161	25	120	281	25	
	August	171	5	120	291	5	
	September	205	10	140	345	10	
	October	255	11	140	395	11	
	November	145	4	150	295	4	Some sources give a figure of 163 Stug III accepted in this month.
	December	24	30	150	174	30	Some sources give a figure of 306 Stug III accepted in this month.
1944	January	77	26	150	227	26	
	February	59	54	137	196	54	
	March	144	56	120	264	56	Conversion of 57 Pzkpfw III chassis begins at MIAG.
	April	194	58	100	294	58	
	May	255	46	80	335	46	
	June	196	100	145	341	100	
	July	242	92	135	377	92	
	August	232	110	80	312	110	
	September	265	119	100	356	119	
	October	253	100	72	325	100	
	November	251	102	110	361	102	
	December	361	40	91	452	40	

The production figures shown above are largely based on those given in Walter J. Spielberger's Sturmgeschutz III & Its Variants *and although this work was published some time ago I have no reason to doubt its accuracy as it is closely aligned with the number of vehicles accepted by the HWA. But it does not list the MAN-built chassis nor is there any mention of the 100 MIAG Pzkpfw III chassis, which would have probably been in the range Fgst.Nr.76361 to 76528, that were converted between April and December 1943 and I have been unable to locate any accurate information on the fate of these conversions.*

Fully-assembled StuG III and StuH 42 vehicles lined up at the Alkett assembly plant in Berlin. The single-piece armour plate bolted in place over the driver's visor and the absence of both the armoured deflector on the commander's cupola and Zimmerit would suggest that this photograph was taken at some time between April and September 1943.

STURMGESCHÜTZ-ABTEILUNG, JUNE 1944

All units of the German Army were organised according to instructions issued by Oberkommando des Heeres (OKH), the high command of the army. These were accompanied by detailed charts referred to as Kriegsstärkenachweisung (KstN) which gave the authorised strength and composition of a unit listing the exact number of personnel, type of vehicle and weapons. They were not issued on a regular basis, as general orders were, but whenever an organisational change was required. Each KstN was numbered and dated, the latter being significant as some numbers were retained when an establishment was upgraded. The battalion depicted here is based around the 10-gun battery, or 31-gun battalion, as most units were never able to carry out the reorganisation to a 45-gun establishment proposed earlier in the year. The additional vehicles needed to implement the increase are shown in a lighter shade. The name changes which affected these battalions is explained in detail on the following page.

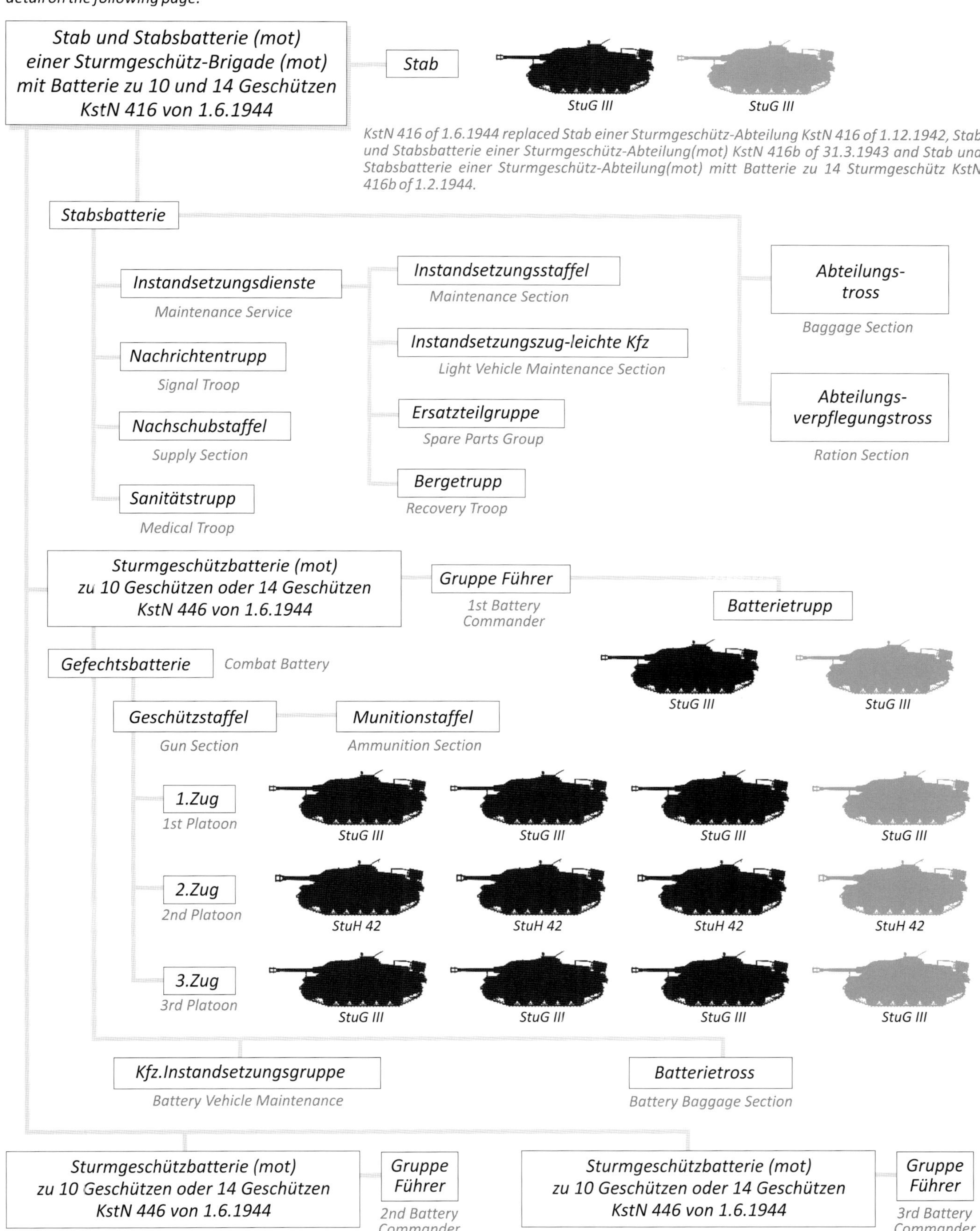

KstN 446 of 1.6.1944 replaced Sturmgeschützbatterie (mot) zu 10 Geschützen KstN 446a of 1.11.1942 and Sturmgeschützbatterie (mot) zu 14 Geschützen KstN 446b of 1.2.1944.

HEERES-STURMARTILLERIE-BRIGADE, 1944

In February 1944 OKH ordered that Sturmgeschütz-Abteilungen were to be renamed Heeres-Sturmgeschütz-Brigaden but this change did not alter the composition of any units and was essentially a means of differentiating the artillery's battalions from the companies of assault guns that had been raised for the Panzerjäger formations of infantry divisions. These companies were henceforth to be referred to, somewhat confusingly, as Sturmgeschütz-Abteilungen but they were uniquely numbered within the division by adding 1000 to the division's number. For example, the assault gun company of 329.Infanterie-Division was referred to as Sturmgeschütz-Abteilung 1329. There were, of course, exceptions to this rule and they are mentioned in the unit histories. In the same month, two new organisational changes were authorised for the independent assault gun battalions which were intended to provide each with a Sturmgeschütz-Begleitbatterie, made up of motorised infantry, and a Begleitpanzer-batterie equipped with Panzer II tanks. Subject to the availability of personnel and equipment, it was anticipated that deliveries would begin in August 1944 and all independent assault gun units would eventually contain these formations. Assault gun battalions which received the Begleit companies were to be renamed Heeres-Sturmartillerie-Brigaden. Only a handful of battalions received these additional units and they are mentioned in the unit histories.

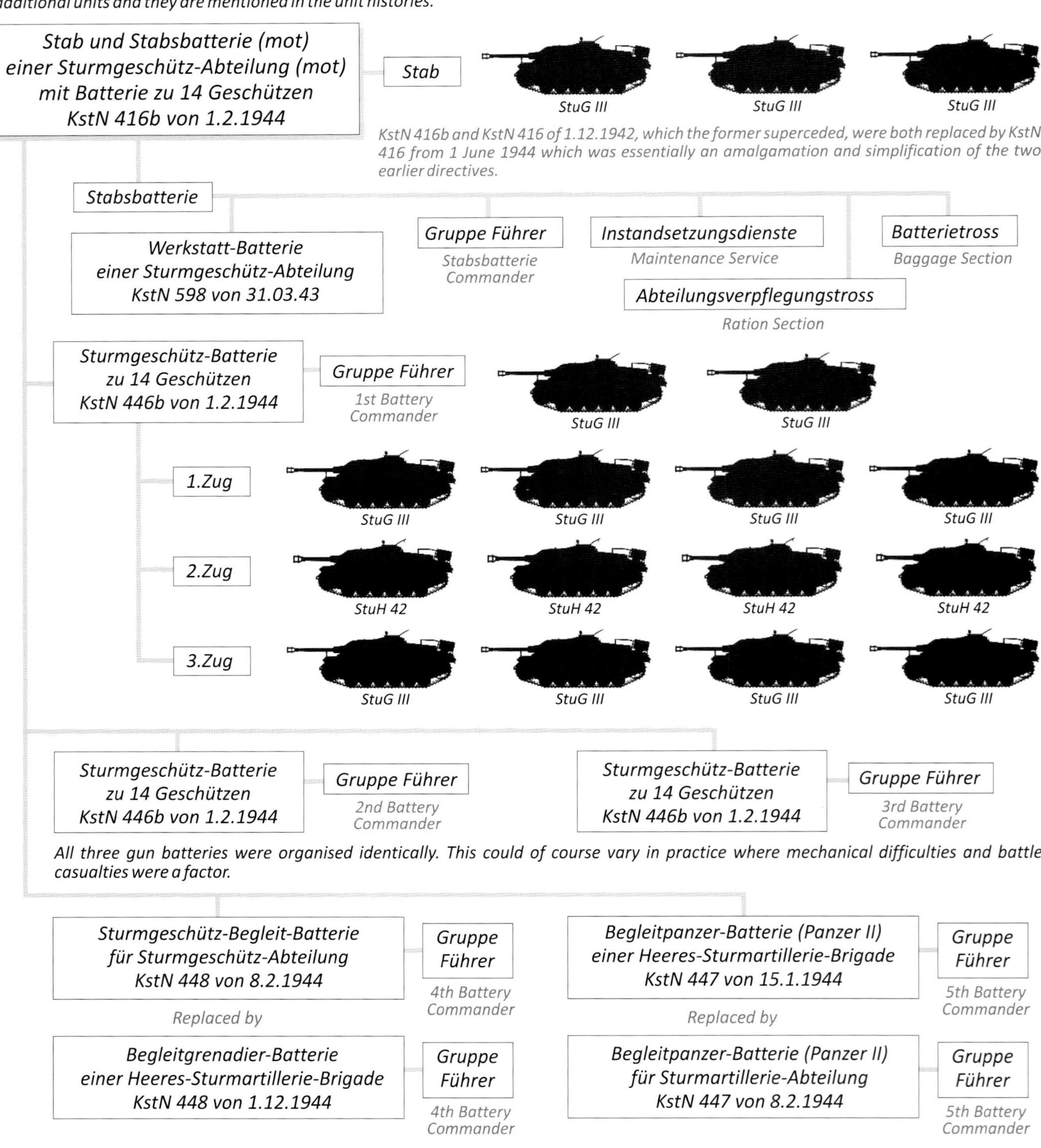

Unfortunately no copy of the initial table of organisation for the Sturmgeschütz-Begleitbatterie, KstN 448 of 8 February 1944, seems to have survived the war. But a proposed change to the battery's establishment put forward on 21 June 1944, which would have allowed for seventeen armoured halftracks, suggests that none were authorised despite the appeals of the artillery inspectorate. In November the motorised infantry company was renamed Begleitgrenadier-Batterie and KstN 448 of 1 December 1944, confirms that the authorised transport was limited to cross country cars and trucks. Each Panzer-Begleitbatterie, organised according to KstN 447 of 8 February 1944, was to have fourteen Pzkpfw II tanks but only four of these units were ever formed and all the vehicles were taken from training units and were in a very poor state. Sturmgeschütz-Brigaden 236 and 239 each received a single battery while Heeres-Sturmartillerie-Brigade 667 was allocated two, which served as the battalion's fifth and sixth batteries. Both were lost in Russia before the battalion was transferred to the West and rebuilt in August and September 1944. All Begleitpanzer-Batterien were officially withdrawn in November 1944 although this order seems to have been largely ignored.

PANZER-ABTEILUNG (FUNKLENK) 302, SUMMER 1944

In November 1939 the firm of Borgward was awarded a contract to develop a fully-tracked, radio-controlled vehicle that could be used to tow mine detonating rollers. Although production versions were taken into service with Minenraum-Abteilung 1 from April 1940, it is doubtful that any were employed in combat. In 1941, in a departure from the roller system of mine clearing, Borgward's purpose-built ammunition carrier was converted to hold a 500kg demolition charge and the resulting vehicle was referred to as the Sdkfz 301 Borgward B IV Sprengladungsträger, usually abbreviated to B IV. At 3.5 tons it was more than twice as heavy as its predecessors and was powered by a 6 cylinder, 49hp engine. The B IV was driven to the front line after which a specially converted tank could control it remotely to its target. Here the demolition charge was jettisoned and, if all went well, the vehicle was recovered by the control tank which also detonated the charge. They were used with some success in late 1942 and at Kursk in July 1943 where, for the first time, they were controlled from specially adapted Sturmgeschütz III assault guns. The formation shown here, Panzer-Abteilung (Funklenk) 302, was formed in June 1944 by drawing together four independent Funklenk (Fkl) companies. The battalion's first company was formed from Panzer-Kompanie (Fkl) 316, its second from Panzer-Kompanie (Fkl) 315, the third from Panzer-Kompanie (Fkl) 317 and the fourth company from Panzer-Kompanie (Fkl) 311. I have assumed the KstN for 1 June 1944 although the companies may have been originally formed using the 1 February 1943 orders. In any case the number of armoured vehicles was the same. Assault guns of this battalion are depicted on page 26 of the Camouflage & Markings section of this book.

Stab, Panzer-Abteilung ausf f (freie Gliederung)
KstN 1107f (fG) von 1.6.1944

Werkstatt-Zug ausf f
KstN 1185f von 1.6.1944

Stabskompanie,
Panzer-Abteilung ausf f (freie Gliederung)
KstN 1150f (fG) von 1.6.1944

Not shown here are the support elements normally associated with the Stabskompanie.

1.Kompanie as
leichte Panzer-Kompanie ausf f
KstN 1171 von 1.6.1944

2.Zug — *As for 1.Zug*

Kfz.Instandsetzungsgruppe

Gefechtstross
Including Sondergerät-Reserve with 12 Borward B IV vehicles

Gepäcktross

It appears that the June 1944 KstN differed little from the version authorised in February 1943, retaining each company's support elements and KstN incorporating the freie Gliederung changes were not actually issued until October 1944. Note that as troops under the control of the Inspekteur der Panzertruppen these units are referred to as Kompanie and not Batterie. The companies were organised identically

2.Kompanie as
leichte Panzer-Kompanie ausf f
KstN 1171 von 1.6.1944

3.Kompanie as
leichte Panzer-Kompanie ausf f
KstN 1171 von 1.6.1944

4.Kompanie as
leichte Panzer-Kompanie ausf f
KstN 1171 von 1.6.1944

The other major unit to take part in the fighting on the Eastern Front late in the war was Panzer-Abteilung (Funklenk) 303 which was formed from the independent Panzer-Kompanie (Fkl) 319, Panzer-Kompanie (Fkl) 301 and Panzer-Kompanie (Fkl) 302 in January 1945. It was soon apparent that the required demolition vehicles would not be available and the battalion was eventually issued with thirty-one Pzkpfw IV tanks and renamed Panzer-Abteilung Schlesien. The last demolition unit to be equipped with assault guns which fought in the East was Panzer-Zug (Fkl) 303 which was scraped together using the surplus personnel and vehicles of Panzer-Abteilung (Fkl) 303 and Panzer-Versuchs und Ausbildung-Abteilung 301. In February 1945, the platoon was issued with four Sturmgeschütz III assault guns and twelve B IV demolition vehicles and attached to 35.Infanterie-Division.

Dragon Models Ltd
B1-10/F., 603-609 Castle Peak Rd,
Kong Nam Industrial Building,
Tsuen Wan, N. T., Hong Kong
www.dragon-models.com

Tamiya Inc
Shizuoka City, Japan
www.tamiya.com

Trumpeter/Hobby Boss
NanLong Industrial Park, SanXiang,
Zhong Shan, GuangDong, P.R.China
www.trumpeter-china.com
www.hobbyboss.com

Academy Plastic Models
521-1, Yonghyeon-dong, Uijeongbu-si,
Gyeonggi-do, Korea
www.academy.co.kr

Hobby Fan/ AFV Club
6F., No.183, Sec. 1, Datong Rd, Xizhi City,
Taipei County 221, Taiwan
www.hobbyfan.com

Black Dog
Petr Polanka
Letecká 549, Libèice nad Vltavou
252 66, Czech Republic
www.blackdog.cz

Italeri S.p.A.
via Pradazzo 6/b,
40012 Calderara di Reno, Bologna, Italy
www.italeri.com

Takom
www.takom-world.com

Border Model
Xizhang Industrial Park, Huishan District,
Wuxi City, P.R. China
www.bordermodel.com

Sturmpanzer.com
www.sturmpanzer.com

Rye Field Models
www.ryefield-model.com
An almost non-existent website. I would recommend one of the on-line retailers.

Hauler
Jan Sobotka,
Moravská 38, 620 00 Brno,
Czech Republic
www.hauler.cz

Voyager
Room 501, No.411 4th Village,
SPC Jinshan District, Shanghai 200540
P.R.China
www.voyagermodel.com

Griffon Model
Suite 501, Bldg 01, 418 Middle Longpan Rd,
Nanjing, P.R. China
www.griffonmodel.com

Aber
ul. Jalowcowa 15, 40-750 Katowice, Poland
www.aber.net.pl

E.T. Model
www.etmodeller.com

Friulmodel
H 8142. Urhida, Nefelejcs u. 2., Hungary
www.friulmodel.hu

Modelkasten
Chiyoda-ku Kanda, Nishiki-Cho 1-7, Tokyo,
Japan
www.modelkasten.com
Very difficult to navigate but worthwhile.

Theodoros Kalamatas Modelling Worshop
www.facebook.com/Theodoros-Kalamatas-Modelling-Workshop

Eduard Model Accessories
Mirova 170, 435 21 Obrnice,
Czech Republic
www.eduard.com

Master Club
www.masterclub.ru
It appears that this firm is closely associated with Armour35, a Russian mail-order firm.

Minicraft Models
I learnt as this book went to print that the Minicraft range will in future be released by Academy Plastic Models.

Passion Models
www.passionmodels.jp

Kaizen
Like many Chinese companies this is hard to pin down to an exact location. The tracks are available from MS Models Japan, a firm for which I can personally vouch.
www.msmodelswebshop.jp

Raupen Model
www.raupen-modell.com
RB Model
Powstancow Wlkp.29B,
64-360 Zbaszyn,
Poland
www.rbmodel.com

M Workshop Singapore
91 Bencoolen St, Sunshine Plaza 01-58,
Singapore
www.themworkshop.com

Royal Model
Via E. Montale, 19-95030 Pedara, Italy
www.royalmodel.com

In writing this book I referred extensively to the *Panzertracts* series of booklets created by the late Thomas Jentz and Hilary Doyle and to a lesser extent, Thomas Anderson's *Sturmartillerie: Spearhead of the Infantry* and *Sturmgeschütz: Panzer, Panzerjäger, Waffen-SS and Luftwaffe Units 1943-45*. The information on unit allocations was all taken for the most part from original Zufürungliste, or supply lists, compiled by the Heereszeugamt between May 1943 and April 1945. Copies of these lists, and many other original documents, can be obtained quite inexpensively from Sturmpanzer.com and I would urge readers to avail themselves of this service. I also relied heavily on the work of Martin Block and the late Ron Klages whose research on vehicle allocations, including those which actually arrived at their destination, would fill many volumes. I would also like to thank Lisa Hooson and Stephen Chumbley, my editors at Pen & Sword, and the very talented modellers who allowed me to include their work. As always, I am indebted to Karl Berne, Valeri Polokov and J.Howard Parker for their invaluable assistance with the photographs and period insignia.

An Alkett-built Sturmgeschütz III of Sturmgeschütz-Brigade 303 photographed during the battle for the town of Ihantala in Finland during the summer of 1944. This vehicle is depicted in the Camouflage & Markings section on page 21. Note what appears to be a non-standard gun support on the hull front.